10/17/89

D0108934

Elastic mechanisms in animal movement

Elastic mechanisms in

animal movement

R. McNEILL ALEXANDER
Professor of Zoology, University of Leeds

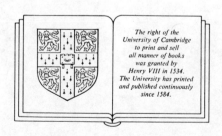

The right of the
University of Cambridge
to print and sell
all manner of books
was granted by
Henry VIII in 1534.
The University has printed
and published continuously
since 1584.

CAMBRIDGE UNIVERSITY PRESS
Cambridge
New York   New Rochelle   Melbourne   Sydney

Published by the Press Syndicate of the University of Cambridge
The Pitt Building, Trumpington Street, Cambridge CB2 1RP
32 East 57th Street, New York, NY 10022, USA
10 Stamford Road, Oakleigh, Melbourne 3166, Australia

First published 1988

Printed in Great Britain at the University Press, Cambridge

British Library cataloguing in publication data

Alexander, R. McNeill
Elastic mechanisms in animal movement.
1. Animal locomotion
I. Title
591,1'852   QP301

Library of Congress cataloguing in publication data

Alexander, R. McNeill.
Elastic mechanisms in animal movement/R. McNeill Alexander.
    p. cm.
Bibliography: p.
Includes index..
ISBN 0 521 34160 4.
1. Animal mechanisms. 2. Elasticity. 3. Animal locomotion.
4. Physiology, Comparative. I. Title.
QP303.A573 1988   87–25661 CIP
591.1'852–dc19

ISBN 0 521 34160 4

PN

Contents

Preface

Springs are useful for many purposes: you can fix a spring on a door, to close it; you can bounce along on the spring of a pogo stick; you can use springs to make a catapult, or the suspension system of a car. Animals exploit the elastic properties of parts of their bodies in ways like these, and in other ways. They use elastic mechanisms in running, jumping, flight, swimming, breathing and in controlling their hands. The study of elastic mechanisms has been a dominant theme in biomechanics at least since the discovery of the protein resilin in 1960 by the late Torkel Weis-Fogh. It has been a remarkably fruitful field of enquiry. We have learned a great deal about animals (including people) and we have had a lot of fun.

This seems to be the first book about elastic mechanisms in animals. I have written it mainly for university students and research workers in biology, but I hope that other people will read it too. They will need some basic understanding of biology, physics and mathematics. I have tried to keep the mathematics simple, and I have avoided skipping 'obvious' steps in the argument. There are a lot of equations, but I think you will find that each leads easily to the next.

I have not tried to include every investigation of elastic mechanisms in animals. Instead, I have selected subjects for their interest and variety, and for their significance for our understanding of animal lives and movements.

R. McNeill Alexander

1

Elastic materials

1.1 Definitions

Elasticity is the property that makes objects spring back to their original shape, after being deformed. A rubber band can be stretched, but when it is released it springs back to its original length. A rubber cushion is squashed flat when I sit on it, but springs back to its original shape when I get up. A bow can be bent, but straightens again when released. Elasticity is most obviously a property of materials like rubber, which can be deformed considerably without breaking, but it is a property of all solids. Steel is elastic, and so also is concrete.

Later chapters in this book are about the functions of elastic materials in the bodies of animals. They show the many ways in which elastic materials are used as springs. This chapter is about the properties of materials, rather than about their functions. A few materials that appear again in later chapters will be used as examples, illustrating some of the methods used to investigate elastic properties and some of the varied properties that are found.

Some terms will be defined by describing an imaginary specimen which (unlike real biological specimens) is perfectly regular and behaves in an ideal way. This ideal specimen is initially a cylinder of length l_0 and uniform cross-sectional area A_0 (Fig. 1.1(a)). It is stretched by forces F, acting on its ends, to a new length l (Fig. 1.1(b)). As it stretches it gets thinner, and the cross-sectional area of the stretched specimen is A.

The greater the force, the more the specimen stretches. This ideal material obeys Hooke's law, which says that the force is proportional to the extension: a graph of F against $(l - l_0)$ is a straight line through the origin (Fig. 1.1(c)). The gradient of the graph is called the tensile stiffness of the specimen (S) and its reciprocal is the tensile compliance (C).

$$S = F/(l - l_0) = 1/C \tag{1.1}$$

Obviously, a given force will stretch a long specimen more than a short one of the same cross-sectional area, made of the same material. It will stretch a slender specimen more than a stout one of the same initial length. Account is taken of effects like these by calculating stress (force/area) and strain (extension/length).

Here we have to be precise, because stress and strain can each be defined in two different ways:

$$
\left.
\begin{array}{ll}
\text{Nominal or engineering tensile stress} & \sigma = F/A_0 \\
\text{True tensile stress} & \sigma_t = F/A \\
\text{Nominal or engineering tensile strain} & \varepsilon = (l - l_0)/l_0 \\
\text{True tensile strain} & \varepsilon_t = \log_e (l/l_0)
\end{array}
\right\} \tag{1.2}
$$

True stress is truer than nominal stress in a fairly obvious way: the force is divided by the cross-sectional area through which it actually

Fig. 1.1.(a) A cylindrical specimen which is stretched, in (b), by a force F. (c) A schematic graph showing the force plotted against the extension. The stippled area represents the strain energy stored in the stretched specimen.

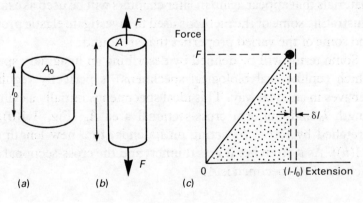

acts. Similarly, a rather tortuous mathematical argument can make true strain seem truer than nominal strain. However, the nominal values are easier to measure or calculate, and are usually used. Many materials can be stretched only a few per cent or less and for them the nominal and true values are almost identical.

Forces are measured in newtons (N) so stresses are expressed in newtons per square metre, also called pascals (Pa). Strains are ratios of two lengths and so are dimensionless, without units. Young's modulus (E) is tensile stress divided by tensile strain, and is measured in pascals. It is a property of the material, not of the particular structure.

$$E = \sigma/\varepsilon \qquad (1.3)$$

Stretching is only one of the ways in which a specimen can be deformed. Another is compression, in which the stress and strain are both negative and can be used as before to calculate Young's modulus. Yet another is shear, the deformation that turns a rectangle into a parallelogram. Fig. 1.2(a) represents a thin rectangular slab of material, with a face of area A glued to a rigid wall. In Fig. 1.2(b) a force F has been applied to the opposite face, parallel to the wall, and has sheared the slab through an angle θ. This angle (expressed in radians) is the shear strain. The shear stress is F/A. The shear modulus of the material is the shear stress divided by the shear strain, and is one-third to one-half of Young's modulus. (It is one-third of Young's modulus for materials that do not change their volume when they are stretched.)

A specimen that is stretched in one direction is sheared in another.

Fig. 1.2.(a), (b) Diagrams illustrating shearing. (c), (d) Diagrams showing that stretching is accompanied by shear. Further explanation is given in the text.

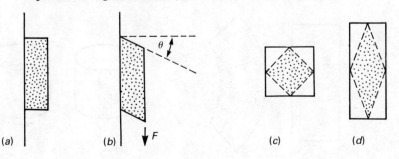

(a) (b) F (c) (d)

Fig. 1.2(c) and (d) represent an initially square block of material that is stretched into a rectangle. The block becomes a rectangle, but the stippled square drawn diagonally on it becomes a rhombus. This shows that stretching the block in a vertical direction involves shearing at 45°.

Bending and twisting are other important kinds of deformation. Fig. 1.3(a) shows that bending involves extension on the outside of the bend and compression on the inside. Fig. 1.3(b) shows that twisting involves shearing around the circumference of the specimen. Mathematical analysis is more complicated for bending and twisting than for stretching and shearing. Other books explain how the stresses and strains can be calculated (for example, Alexander, 1983; Currey, 1984).

A stretched spring or a bent bow has energy (called elastic strain energy) stored up in it. Work is needed to stretch the spring or bend the bow, to supply the strain energy, and most of this work can be recovered in an elastic recoil. In the case of the specimen shown in Fig. 1.1, a force F stretches it to length l. The work needed to stretch the specimen a small additional amount δl (too small to involve an appreciable increase of force) is $F \cdot \delta l$: remember that work is force multiplied by the distance its point of application moves along its line of action. This work $F \cdot \delta l$ is also the area of the narrow strip indicated by the vertical broken lines in Fig. 1.1(c). Thus, the work done as the force increases from zero to F is the stippled area under the graph, which is equal to $\frac{1}{2}F(l - l_0)$. This work is converted to strain energy U:

$$U = \tfrac{1}{2}F(l - l_0) \tag{1.4}$$

Fig. 1.3. Diagrams showing (a) that bending involves stretching and compression, and (b) that twisting involves shear.

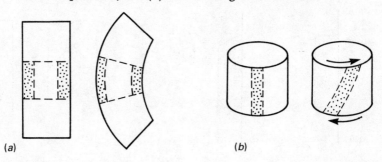

(a)   (b)

By using equation 1.1 to substitute for F or $(l - l_0)$, we can get other useful forms of the equation:

$$U = \tfrac{1}{2}F^2/S = \tfrac{1}{2}S(l - l_0)^2 \tag{1.5}$$

Alternatively, equation 1.4 can be converted to a different form by using definitions of nominal stress and strain from equation 1.2:

$$U = \tfrac{1}{2}\sigma A_0 \varepsilon l_0 = \tfrac{1}{2}\sigma \varepsilon V \tag{1.6}$$

where $V (=A_0 l_0)$ is the volume of the specimen.

Most of the remaining sections of this chapter are about the properties of a few examples of biological materials. A good deal of practical detail will be given to show how the properties have been discovered.

1.2 Ligamentum nuchae: properties

The ligamentum nuchae is a highly extensible ligament, composed largely of the protein elastin. It runs along the backs of the necks of hoofed mammals, and helps to support the weight of the head. Its function is discussed in Chapter 2, and it appears here only as an example of a structure whose elastic properties can be investigated extremely easily. Dimery, Alexander & Deyst (1985) performed very simple experiments on the ligamentum nuchae of sheep and deer.

The ligament was dissected out, leaving it attached to the back of the skull. The head was detached from the body and fastened to a firm support, with the ligament hanging down. A pan was tied to the lower end of the ligament so that weights could be put in it to stretch the ligament (Fig. 1.4(a)). Care was taken to keep the specimen moist: the properties of biological specimens generally change, if they are allowed to dry.

The ligament tapers (it is most slender near the head) so any load sets up different stresses and strains in its different parts. This difficulty was overcome by sticking pins through the ligament, and measuring the distances between successive pins for each load. Thus, the strain for each segment (from one pin to the next) could be calculated. Afterwards the specimen was cut into segments and the cross-sectional area of each determined, so that stresses could be calculated for each load.

A segment of mass m and density ϱ has volume m/ϱ. Its cross-sectional area is this volume divided by its length l: it is $m/\varrho l$. Thus, cross-sectional areas of ligaments can be calculated from their masses and lengths. This method of measuring cross-sectional areas is often convenient to biologists, but in this particular investigation the volume was measured more directly. A small beaker containing water was placed on a balance, and the reading noted. The segment of ligament, suspended by a fine thread, was lowered into the beaker until it was submerged without touching the glass. This made the reading of the balance increase by an amount equal to the weight of water displaced (by Archimedes' principle). The density of water is 1.00 g cm^{-3}, so an increase of x grammes indicated a volume of x cubic centimetres.

Thus, graphs of stress against strain were obtained (Fig. 1.4(b)). If Hooke's law were obeyed, the graph would be straight, with gradient equal to Young's modulus. However, the graph curves. In cases like this it is customary to measure tangent Young's moduli, which are the gradients of tangents to the curve. Fig. 1.4(b) gives a tangent

Fig. 1.4.(a) A diagram of an experiment on the ligamentum nuchae of a deer (*Capreolus capreolus*). (b) A graph of stress against strain, obtained from the experiment. The gradient of the tangent is the tangent Young's modulus at a stress of 0.5 MPa. The data are from the work of Dimery, Alexander & Deyst (1985).

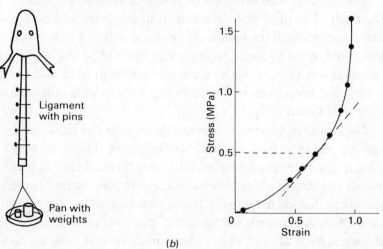

modulus of 1.2 MPa, at a stress of 0.5 MPa, and higher moduli at higher stresses.

1.3 Tendon

Tendon consists largely of the protein collagen. It is much less extensible than ligamentum nuchae: it breaks at strains of about 0.08 (data of Bennett, Ker, Dimery & Alexander, 1986), whereas ligamentum nuchae can be stretched to strains of about 1. This makes it harder to make acceptably accurate measurements of its elastic properties. The experiments on ligamentum nuchae, described in the previous section, were crude but adequate. The experiments on tendon described in this one were much more sophisticated and required immensely more expensive equipment (Ker, 1981).

Tendons were stretched in a dynamic testing machine (Fig. 1.5(a)). A length of tendon is held at its ends by clamps. The lower clamp is attached to a hydraulically driven actuator which can

Fig. 1.5. Diagrams of (a) a dynamic testing machine, and (b) an extensometer.

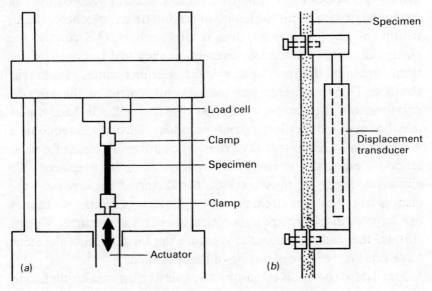

be made to move down and up as required, stretching the tendon and allowing it to recoil. The upper clamp is attached to a load cell which senses the force on the tendon. Electrical outputs from the machine indicate force and actuator position and can be used to produce records like Fig. 1.6(*a*).

Notice that the record is a loop. The upper line was drawn as the specimen was stretched and the lower line as it shortened. The loop is formed because some of the work done stretching the specimen is degraded to heat instead of being recovered in the recoil. This happens to some extent with all materials but was not apparent in the crude tests on ligamentum nuchae (section 1.2).

The area $(A + B)$ under the rising line represents the work done stretching the specimen, just as the stippled area in Fig. 1.1(*c*) does. The area B represents the work recovered in the elastic recoil and the loop area (A) the energy lost as heat. There are several confusingly different ways of describing the proportion of energy lost (Ker, 1981). The one used in this book is the energy dissipation $A/(A + B)$. Another one that is commonly used by engineers is the loss tangent, which is about 0.6 times the energy dissipation if both are small.

Fig. 1.6(*a*) was obtained from expensive equipment, but is nevertheless ambiguous and potentially misleading. First, each clamp gripped an appreciable length of tendon (about 20 mm). This makes it impossible to calculate strains accurately, because we cannot be precise about the length of tendon being stretched: where, within the gripped regions, does it effectively end? Secondly, the clamps distorted the tendon severely (as they had to do, to grip it firmly enough). Different parts of the specimen suffered different stresses. Thirdly, energy was used by movement of the gripped region within the clamps, as tension rose and fell. All these effects may be small for very long, slender tendons, such as the tendons of kangaroo tails (Bennett *et al.*, 1986). They are troublesome for most tendons, especially if energy dissipation is to be measured. To eliminate them, we must exclude the distorted regions near the clamps from our measurements of length change. Thus, we cannot use actuator displacement as our measure of length change. We can still use the load cell output to measure the force because the same force acts on every cross-section of the specimen.

Fig. 1.5(*b*) shows an extensometer, a device for measuring length changes in the undistorted part of the specimen (Ker, 1981). It grips

Fig. 1.6. Graphs of force against displacement, obtained when the gastrocnemius tendon of a wallaby (*Macropus rufogriseus*) was tested as illustrated in Fig. 1.5. The actuator moved up and down sinusoidally, with a frequency of 2.2 Hz. In (*a*) force and displacement were both obtained from the outputs of the testing machine. In (*b*) length changes of a shorter segment of the tendon were measured by means of an extensometer. Additional scales show stress and strain. From the data of Ker, Dimery & Alexander (1986).

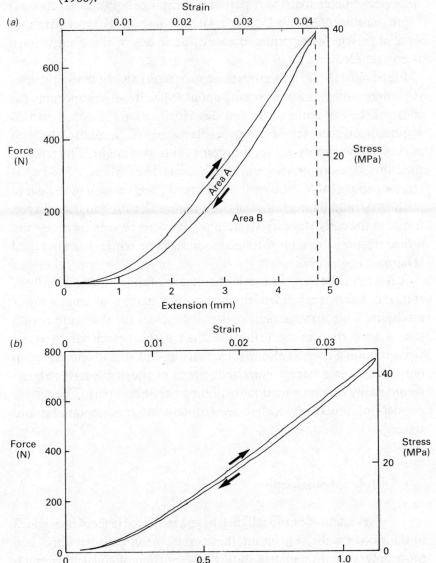

two points on the tendon lightly, and gives an electrical output that indicates the distance between them. A light grip, which barely distorts the tendon, is adequate because the grips do not have to transmit large forces: they need only be firm enough to overcome the slight friction and inertia of the extensometer.

Fig. 1.6(*b*) shows a record of a test using the extensometer, on the same tendon as Fig. 1.6(*a*). The length changes are smaller because the extensometer spans only part of the distance between the clamps. More significantly, the loop is relatively narrower, representing a smaller energy dissipation, because the losses in the clamps have been excluded.

Fig. 1.6(*b*) shows two interesting characteristics of tendon. First, the energy dissipation is small, about 0.07. This is important for some of the functions of tendon described in later chapters, which require good recovery of energy in elastic recoil. Secondly, the loop curves at low strains but soon becomes almost straight. The tangent modulus is more or less constant (about 1500 MPa, 1.5 GPa) at stresses above 30 MPa. The initial curved ('toe') region of the curve is due to straightening out of slight waviness ('crimp') in the collagen fibrils in the early stages of stretching. This can be seen by observing a fine strand of tendon through a microscope while it is stretched (Diamant *et al.*, 1972).

A tendon is not a solid block of collagen. Rather, it is a bundle of collagen fibres embedded in a jelly-like matrix of much lower modulus. This arrangement makes it flexible, for the same reason that a wire rope is more flexible than a solid steel bar of equal cross-sectional area. It also makes it strong and stiff in one direction only. Transverse forces (which tendons do not normally have to bear) would easily disrupt a tendon by pulling the fibres apart. The matrix consists of mucopolysaccharides (protein–sugar compounds) and water.

1.4   Mesogloea

Sea anemones and jellyfish have a jelly-like layer of mesogloea in their body walls. This is another composite of collagen fibres in a mucopolysaccharide matrix, but the proportion of collagen is much less than in tendon and the fibres run in all directions.

Gosline (1971) cut strips of mesogloea from the body walls of sea anemones (*Metridium*) and investigated their elastic properties. To do this, he built a miniature dynamic testing machine, sensitive enough to record very small forces. Clamping difficulties were avoided by gluing the specimen to the machine. He made dynamic tensile tests at various frequencies and found, for example, that at 11°C Young's modulus was 100 kPa in tests at 1.5 Hz, but only 60 kPa at 0.01 Hz. Low frequencies allowed more time for the specimen to stretch, in each cycle, and it stretched more.

He also made tests of another kind, using the same machine. He stretched the specimen suddenly by 20–30% and then held it at its new length. The force gradually declined, a phenomenon called stress relaxation. He calculated Young's modulus at successive times by dividing the stress at the time in question by the (constant) strain. In an experiment at 13 °C, for example, he found that the modulus fell from 20 kPa at 10 s to 0.2 kPa at 10^5 s (28 h). Very little change of stress occurred in the last 20 h of the experiment: the stress seemed to be approaching an equilibrium value.

Alexander (1962) and Koehl (1977) made yet another kind of test. Instead of applying a strain and observing the changing stress, they applied a constant stress and observed the changing strain. The specimen stretched gradually over a long period of time, a phenomenon called creep (Fig. 1.7). As in Gosline's experiments, *Metridium*

Fig. 1.7. Graphs of (nominal) strain against time for strips of sea anemone mesogloea, stretched by constant (true) stresses of about 3 kPa. Results for two species are shown. Redrawn from Koehl (1977).

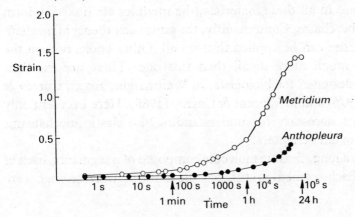

mesogloea approached equilibrium after times of the order of 10^5 s. If the stress was removed after such times, the specimen returned gradually to its initial length. Plainly, the length changes, though slow, were elastic.

Other biological materials, including ligamentum nuchae and tendon, also show creep and stress relaxation, and their moduli in dynamic tests depend on the frequency. Mesogloea is not peculiar in showing these time- and frequency-dependent properties, but it shows them very clearly.

1.5   Molecular mechanisms

Sections 1.2 to 1.4 should have given some impression of the variety of mechanical properties shown by the materials discussed in this book, and of the techniques that are used for investigating them. Many more materials will be introduced later, some of them with properties well outside the range illustrated by ligamentum nuchae, tendon and mesogloea. Nevertheless, it seems useful at this stage to give some explanation of the molecular arrangements that give the materials their varied properties.

All the materials whose elastic properties are discussed in this book are high polymers. That means that their molecules are composed of large numbers of similar modules. In proteins, the modules are amino acid residues. In polysaccharides, they are monosaccharide units. Similarly, rubbers and plastics are high polymers: rubber molecules are built from numerous isoprene modules and polystyrene (as its name suggests) is built from modules of the hydrocarbon styrene. In all these materials, the modules are linked to form long, flexible chains. Consequently, the same basic theory of mechanical properties can be applied to them all. Other books present the theory in much more detail than this one. There are excellent accounts, designed for biologists, in Wainwright, Biggs, Currey & Gosline (1976) and in Vincent & Currey (1980). Here I present only what seems necessary for understanding the elastic mechanisms described in later chapters.

Imagine a long, flexible molecule composed of n segments, each of length a. Each segment is stiff, but the joints between segments are flexible. Similarly, a metal chain consists of stiff links, flexibly joined together. Assume that this molecule is not restrained in any way, but is free to move. It will not remain stationary (unless the temperature is absolute zero) but will writhe in Brownian movement. It will curl and uncurl so that its ends move closer together or further apart. At a particular instant, the distance from one end to the other may have any value from zero (if the ends are in contact) to na (if the molecule is stretched out straight). However, the extreme lengths are unlikely. It can be shown that the most probable length for the molecule is $n^{1/2}a$.

Now imagine a three-dimensional network of such chains, joined together at their ends (Fig. 1.8(a)). This is a model of the molecular structure of materials like rubber and elastin. The process of vulcanizing rubber puts sulphur links between the isoprene chains, connecting the molecules into a network. Similarly, the protein chains of elastin are linked by compounds called desmosines (Gosline, 1980). The connections that join the polymer molecules into a network are called cross-links.

The molecular chains in the network will continue to writhe and the junctions at which they end will move together and apart. The distance between the ends of a particular chain will vary but the average distance, in a sample containing many chains, will always be $n^{1/2}a$. (This is actually a root mean square value, not a simple mean.)

Now suppose an external force acts, to stretch the sample. The network is distorted (Fig. 1.8(b)). Chains aligned in the direction of stretching (for example, chain AB) now have average lengths greater than $n^{1/2}a$, and those aligned across the direction of stretching (BC) have average lengths less than $n^{1/2}a$. The chains are still writhing and a particular chain will sometimes be aligned predominantly along

Fig. 1.8. A diagram of a rubbery polymer (a) unstrained, and (b) stretched horizontally. The sinuous lines represent polymer molecules and the dots represent cross-links.

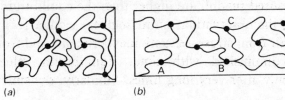

(a)   (b)

the direction of stretching and sometimes across it. However, the arrangement is no longer random. When the distorting force is removed, the writhing chains move back into random configurations and the sample returns to its original shape: it recoils elastically.

A theory developed from these ideas predicts the force required for equilibrium of the distorted network, and hence the modulus of elasticity. The shear modulus G and Young's modulus E are given by

$$G = NkT = E/3 \qquad (1.7)$$

where N is the number of chains per unit volume of material, k is a physical constant (Boltzmann's constant) and T is the absolute temperature. Notice the importance of the number of chains. If more cross-links are inserted, dividing the molecules into a larger number of shorter chains, the material becomes stiffer. Also, the modulus is proportional to the absolute temperature because the energy associated with the writhing of the molecules increases with temperature. Similarly, the pressure exerted at constant volume by a gas increases with temperature, because the kinetic energy of the molecules increases with temperature.

The number of chains per unit volume can be calculated from the mass of polymer per unit volume (ϱ) and the mean molecular mass of the chains (M). This makes it possible to re-write equation 1.7 in the form

$$G = \varrho RT/M = E/3 \qquad (1.8)$$

where R is the universal gas constant ($8.3 \text{ J K}^{-1} \text{ mol}^{-1}$). Notice that ϱ is the mass of polymer per unit volume of material, and may be less than the density of the material if it also contains other substances (for example, water).

The moduli given by equations 1.7 and 1.8 are equilibrium values, calculated from stresses and strains reached at equilibrium. It takes time to reach equilibrium, by random writhing of molecules. Long molecules can get into dreadful tangles, and the times required are sometimes very long. If too little time is allowed for equilibrium to be reached, measured moduli are higher than the equilibrium values. This explains the observations on mesogloea (section 1.4): moduli calculated both from stress–relaxation experiments and from creep experiments diminished with time. Equilibrium values were approached only after very long times. Similarly, dynamic tests at high frequencies allowed less time for equilibration than at low frequencies, and gave higher moduli.

Fig. 1.9 shows how modulus is related to time (or frequency) for a typical amorphous polymer. At very short times (high frequencies), there is virtually no time for equilibration by molecular writhing. The mechanism of elasticity that has been described is ineffective, and the material behaves like an ordinary non-rubbery solid. Young's modulus is very high (typically 2–10 GPa) and the material breaks at very small strains. This is called glassy behaviour. Polystyrene is an example of a polymer that shows glassy behaviour at all reasonable times, at room temperature.

At longer times (lower frequencies), the modulus is lower. Moderate times allow a good deal of writhing, but not enough for entanglements between the molecules to become fully disentangled. These entanglements behave as additional cross-links, keeping the modulus above the equilibrium value. The modulus falls as time passes but the fall may be interrupted, in graphs like Fig. 1.9, by a plateau.

Eventually equilibrium is reached, if the material is cross-linked. The modulus then remains constant, as time increases further. However, if there are no cross-links, the modulus falls eventually to zero (Fig. 1.9, broken line). Polymers with no permanent cross-links can behave elastically because molecular entanglements serve as cross-links, but these eventually disentangle and the molecules flow past each other. Consequently, the material behaves as an elastic solid at short times, but as a viscous liquid at long times. The bouncing putty sold as a toy is a material of this kind.

Fig. 1.9. A schematic graph of the logarithm of Young's modulus against the logarithm of time, for an amorphous cross-linked polymer or (broken line) for one that is not cross-linked.

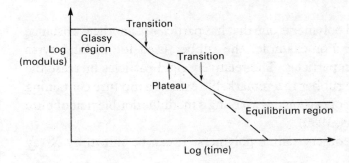

Different materials reach equilibrium at different times. Elastin is close to equilibrium after only 0.1 s (Gosline, 1980) but *Metridium* mesogloea needs about 10^5 s (Fig. 1.7). Compare the curve for *Metridium* in this figure with the schematic graph (Fig. 1.9), remembering that one shows strains and the other moduli. At times below 1 min, the mesogloea has small strain. The modulus is about 30 kPa and the specimen seems to be in a plateau region. At times over 10 h, the modulus seems to be settling at an equilibrium value of about 2 kPa. Between these times, a transition occurs.

Different materials also have very different equilibrium moduli. Soft rubbers, elastin and two other rubbery proteins, abductin and resilin, all have equilibrium Young's moduli of the order of 1 MPa, but mesogloea has a value of about 200 Pa. The difference is due to mesogloea containing a very small concentration of polymer (ϱ is small, equation 1.8) and consequently having few chains per unit volume. However, its modulus can be increased by treating it with formaldehyde (Gosline, 1971). This introduces extra cross-links, dividing the molecules into a larger number of shorter chains: it reduces M (equation 1.8).

The theory that has been presented in this section refers to homogeneous amorphous polymers. Homogeneous means that they have the same composition throughout. Amorphous means that the molecules are not lined up in crystalline arrays. The tissues that have been discussed are not truly homogeneous (mesogloea, in particular, contains collagen fibres) but no theoretical account has yet been taken of this. The next section describes the consequences of inhomogeneity and crystallinity.

1.6   Filled polymers and fibres

A filled polymer is one that has particles of another substance mixed with it. For example, the rubber used for making tyres contains carbon particles. The relatively rigid particles increase the modulus of the rubber to a remarkable extent: a mixture containing 20% by volume of carbon has a Young's modulus double that of pure rubber (Mullins, 1980).

Fibres are semi-crystalline polymers. It can be shown by X-ray diffraction that parts of the polymer molecules are lined up parallel to the length of the fibre, in crystalline arrays. The crystalline regions have a high modulus because the molecules in them are not free to writhe. Consequently, they behave like rigid filler particles. They increase Young's modulus, giving fibres much higher moduli than amorphous polymers. For example, tendon (which consists mainly of collagen fibres) has a modulus of 1.5 GPa (section 1.3). This is much higher than the moduli of elastin, abductin and resilin, which are of the order of 1 MPa.

The relationship between fibres and rubbery polymers is illustrated by the effect of heating tendon above 65 °C. This breaks down the crystalline regions, allowing the molecules to move out of alignment into a random arrangement. The tendon shortens to one-third of its initial length and becomes rubbery, with a much lower Young's modulus.

Many biological materials consist of fibres embedded in a matrix. Mesogloea has some collagen fibres in its jelly-like matrix. They presumably do not run continuously through the matrix (if they did, they would prevent it from stretching much) but act as a filler, increasing its modulus (Gosline, 1971). Tendon consists of collagen fibres in a matrix, as already noted (section 1.3). Insect cuticle consists of chitin fibres in a protein matrix. Composite materials like these, with fibres running predominantly in particular directions, are anisotropic. This means that their mechanical properties depend on the direction of the applied stress. For example, the apodemes of insects are inward extensions of the cuticle to which muscles attach. One studied by Ker (see Vincent, 1980) had all its chitin fibres running parallel to its length. Its Young's modulus was 11 GPa for lengthwise stretching but only 0.15 GPa for transverse stretching.

1.7 Muscle

Muscle fibres are capable of only slight elastic stretching and recoil. Large length changes are not elastic, but are due to protein filaments sliding past each other, much as the segments of a telescope slide past each other when the telescope is lengthened or shortened. The forces that act when the muscle is active are generated by cross-bridges between the filaments. Lengthening or shortening of the active muscle involves cross-bridges detaching, and re-attaching further along.

Small elastic length changes can occur, without detachment of cross-bridges. This has been shown by experiments in which a single fibre from a frog muscle was held at constant length while being stimulated electrically to exert tension (Huxley & Simmons, 1971). Suddenly, the clamps holding the fibre were moved slightly closer together. The tension fell, just as the tension in a stretched rubber band falls when it is allowed to shorten, but the fibre (unlike the rubber band) re-developed tension during the first few milliseconds at its new length. It did this by moving cross-bridges to take up the slack. The tension fell briefly to zero whenever the active fibre was allowed to shorten by 1.3 % or more. This showed that the tension in the active fibre stretched it by 1.3 %.

Huxley & Simmons (1971) obtained evidence that most of the elastic compliance was in the cross-bridges. Their argument is illustrated by Fig. 1.10, in which the cross-bridges are represented by tiny springs. Each attached cross-bridge is stretched by the force produced by itself. It seems reasonable to assume that the force and the extension have the same values for every cross-bridge. The sarcomere shown in Fig. 1.10 has twice as many cross-bridges attached when it is short (*b*) as when it is longer with less overlap between the filaments (*a*). It should exert twice as much force but

Fig. 1.10. Diagrams of striated muscle, to illustrate the discussion of its elastic properties. The horizontal lines represent actin and myosin filaments, and the cross-bridges are represented as springs.

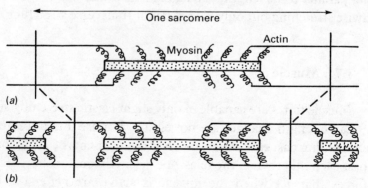

have the same elastic extension. Gordon, Huxley & Julian (1966) had already shown that the force was proportional to the number of attached cross-bridges over a wide range of sarcomere lengths (but not, for reasons they explained, at short lengths). Huxley & Simmons (1971) showed that the elastic extension was the same, about 28 nm per sarcomere, for sarcomere lengths of 2.2 and 3.2 μm (2200 and 3200 nm).

Stimulated muscle fibres exert more force when being forcibly stretched than when held at constant length: fast stretching can double the force. During stretching, each cross-bridge exerts more force, and should stretch itself more. Flitney & Hirst (1978) performed a variant of the Huxley & Simmons experiment, in which stretching of an active frog muscle fibre was interrupted by sudden shortening. They found that elastic extensions up to 60 nm per sarcomere (3 %) occurred in fast stretching. Rack & Westbury (1974) had obtained a similar result from experiments on cat muscles.

Many muscles attach to the skeleton through tendons, so that the elastic compliances of muscle fibres and tendons act in series. In later chapters we will want to know which is more important. The discussion that follows is based on a paper by Alexander & Bennet-Clark (1977). In some muscles, described as parallel-fibred, the fibres are in line with the tendons (Fig. 1.11(a)). In others, described as pennate, the fibres approach the tendons at an angle (Fig. 1.11(b)).

Fig. 1.11. Diagrams of (a) a parallel-fibred muscle, and (b) a pennate muscle. (c) and (d) are explained in the text.

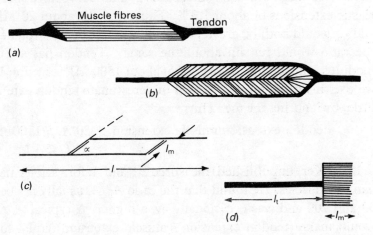

In a parallel-fibred muscle, the force transmitted by the tendon equals the force (F_m) exerted by the muscle fibres. In a pennate muscle it is only $F_m \cos \alpha$, if the fibres make an angle α with the tendon. In a parallel-fibred muscle, the force is transmitted through a length $(l - l_m)$ of tendon, where l is the overall length of muscle plus tendon and l_m is the length of the muscle fibres. In a pennate muscle, each muscle fibre exerts its force through a total length $(l - l_m \cos \alpha)$ of tendon: notice that this is true for both the muscle fibres shown in Fig. 1.11(*c*). However, the angle α is usually less than 30° in muscles of vertebrates, making $\cos \alpha$ almost equal to 1 ($\cos 0 = 1$: $\cos 30° = 0.87$). It will be accurate enough for our purposes to treat all muscles, parallel-fibred or pennate, as exerting force F_m and as consisting of muscle fibres of length l_m in series with a tendon of length $l_t = (l - l_m)$ (Fig. 1.11(*d*)). Let the cross-sectional areas of the muscle and the tendon be A_m and A_t, respectively. A_m means the total of the cross-sectional areas of all the muscle fibres.

We want to know how much the muscle fibres and tendon stretch. Let the muscle fibres exert a stress σ_m. The force F_m is then $A_m \sigma_m$ and the stress in the tendon is $A_m \sigma_m / A_t$. Let the tendon obey Hooke's Law, with Young's modulus E_t. The strain in it is then $A_m \sigma_m / A_t E_t$ and its extension is $A_m \sigma_m l_t / A_t E_t$. Let the elastic strain in the muscle fibres be ε_m so that their elastic extension is $\varepsilon_m l_m$. The ratio (tendon extension)/(muscle extension) is $A_m \sigma_m l_t / A_t E_t \varepsilon_m l_m$.

We can simplify that complicated expression. Vertebrate muscle fibres, contracting isometrically at their optimum lengths, exert stresses of about 0.3 MPa (Close, 1972) and, as we have seen, have elastic extensions of about 1.3%. Thus, σ_m / ε_m is about 20 MPa (σ_m and ε_m would both be greater if the fibres were being stretched, but the ratio would remain about the same). Tendon has a Young's modulus, outside the toe region, of about 1500 MPa (section 1.3). If we use this value for E_t, we will underestimate tendon extension a little, by ignoring the toe. Thus,

$$\text{tendon extension/muscle extension} \simeq 20 \, A_m l_t / 1500 \, A_t l_m$$
$$\simeq A_m l_t / 75 A_t l_m \qquad (1.9)$$

R. F. Ker (unpublished) measured A_m and A_t for various mammalian leg muscles. He found that the ratio A_m / A_t usually lay between 30 and 100 and was occasionally even higher. A typical value, 75, would make (tendon extension)/(muscle extension) equal to l_t / l_m.

This gives us a conveniently simple rule. If a muscle has long fibres and short tendons, most of the elastic extension occurs in the muscle fibres. If it has short muscle fibres and a long tendon, most of the elastic extension is in the tendon.

Strain energy is $\frac{1}{2}$ (force) $\times$ (extension), for a body that obeys Hooke's Law (equation 1.4). The same force acts on the tendon and the muscle fibres. Therefore, a muscle with long fibres and short tendons stores strain energy mainly in the muscle fibres. One with short fibres and a long tendon stores strain energy mainly in the tendon.

2

Springs as muscle antagonists

The most obvious thing that a spring will do is to reverse a movement. If you stretch a spring and release it, it springs back. A spring can be used to make a door close automatically. Similarly, elastic structures can be used to reverse movements caused by muscles.

2.1 Bivalve shells

Clams and other bivalve molluscs have a shell in two parts, hinged together. It can be closed to protect the soft parts of the body. Alternatively, it can be opened to admit the water currents from which the mollusc filters out food particles, or to allow the foot to be protruded for burrowing. It is closed by contraction of a muscle, but it is opened again by a spring.

Fig. 2.1(a) shows the arrangement in the scallop *Pecten*. The two valves are held together at the hinge by a strip of relatively inextensible protein, the outer hinge ligament. Just inside this is the inner hinge ligament, a block of the protein abductin. This is a rubber-like material; if you dissect it out and drop it on the floor, it bounces like rubber. Its Young's modulus is about 4 MPa, which is the same order of magnitude as soft rubber and elastin (Alexander, 1966). When the adductor muscle contracts and closes the shell, the abductin is compressed. When the muscle relaxes, the abductin recoils, opening the shell again. This seems a neat solution to the problem of opening a shell from inside. (Remember that muscles cannot push, they can only pull.) It has, however, the disadvantage that the muscle must be kept taut enough to compress the abductin, for as long as the shell remains closed.

Trueman (1953) used the apparatus shown in Fig. 2.1(*b*) to investigate the properties of bivalve hinge ligaments. The soft tissues had been removed from the shell but the hinge was intact and still moist. The shell closed when weights were added to the pan and sprang open again when they were removed. Trueman observed the angle of opening of the shell with different weights in the pan. He calculated the moments that the apparatus exerted about the shell's hinge, and plotted these moments against the angles of opening (Fig. 2.1(*c*)). Areas in graphs like this represent energies, just as they do in the graphs of force against extension obtained from tensile tests (Fig. 1.6). The narrow loop for *Pecten* in Fig. 2.1(*c*) indicates that energy dissipation in its hinge ligament is small. The wider loops for *Cyprina* (a larger mollusc) and *Mytilus* (a smaller one) indicate larger percentage energy dissipations.

Fig. 2.1.(*a*) Diagrammatic section through a scallop (*Pecten*) showing the hinge ligaments and adductor muscle. (*b*) Apparatus used to investigate the mechanical properties of bivalve mollusc hinge ligaments. (*c*) Results of tests using this apparatus. The applied moment about the hinge is plotted against the angle of opening of the shell, for an incomplete cycle of closing and opening of each of three species. Redrawn from Trueman (1953).

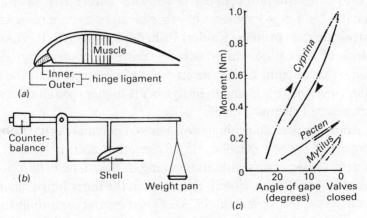

Pecten and other scallops are exceptional among bivalve molluscs, in being able to swim. They open the shell and clap it shut again about three times per second, producing jets of water that lift them off the bottom and propel them at speeds of about 0.3 m s $^{-1}$. This peculiarity seems to explain why they have evolved hinge ligaments with exceptionally low energy dissipations.

Trueman (1953) gave no figures for energy dissipation and his experiments were in any case far too slow to imitate the swimming movements of scallops. Alexander (1966) tested small scallops (*Chlamys*) at a frequency of 3 Hz and showed that the energy dissipation was 9% or less. No measurements have been made on non-swimming bivalves at such high frequencies, but it seems clear that their energy dissipations would be very much larger.

Moore & Trueman (1971) measured the volume of water squirted out by *Chlamys* in each swimming stroke, and the pressure used to squirt it. It can be calculated from their data that the work done on the water in each squirt is about 13 mJ. Trueman's (1953) data show that the work done compressing the hinge ligament, in the same movement, is about 8 mJ. Most of that 8 mJ is recovered in the elastic recoil, when it does the work of opening the shell. However, if the energy dissipation were high, much of the work done compressing the ligament would be lost. Such a loss would considerably increase the energy cost of swimming. The low energy dissipation of scallop ligament gives a substantial advantage.

Scallop hinge ligament is chemically different from the hinge ligaments of other bivalves. The amino acids in its protein (abductin) are in different proportions. It contains hardly any calcium carbonate, but hinge ligaments of other bivalves contain more calcium carbonate than protein (Kahler, Fisher & Sass, 1976). It is uncertain which of the various differences has most effect on energy dissipation. The calcium carbonate in non-swimming molluscs acts as a filler (section 1.6), giving their ligaments higher moduli than scallop ligaments (Trueman, 1953).

Abductin is an amorphous cross-linked polymer, as its rubber-like properties suggest (section 1.5). Like other such polymers, it can undergo large strains without breaking. It has to, to do its job. When the shell of *Pecten* is closed, the strain in the inner hinge ligament is about -0.45 (Trueman, 1953). (The negative sign indicates that this is a compressive strain.)

2.2 Ligamentum nuchae: function

The ligamentum nuchae in the necks of hoofed mammals serves as a tension spring (Fig. 2.2(a)). It runs along the dorsal part of the neck, from the neural spines of the thoracic vertebrae to the skull. It sends branches to some (as in sheep) or all (as in camels) of the neck vertebrae. When the head is lowered, for example for feeding, the ligament is stretched (Fig. 2.2(b)). When the head is raised, the ligament recoils elastically. Dimery *et al.* (1985) measured the strains in the ligament at different head positions. We dissected

Fig. 2.2.(a),(b) Outlines traced from X-ray pictures of a roe deer (*Capreolus capreolus*) carcase with the head in the alert position and lowered for feeding. The position of the ligamentum nuchae is indicated by stipple. From the data of Dimery, Alexander & Deyst (1985). (c) Outline traced from an X-ray picture of the neck of a turkey (*Meleagris gallopavo*). The elastic ligaments of the proximal part of the neck are indicated by stipple. From the data of Bennett & Alexander (1987).

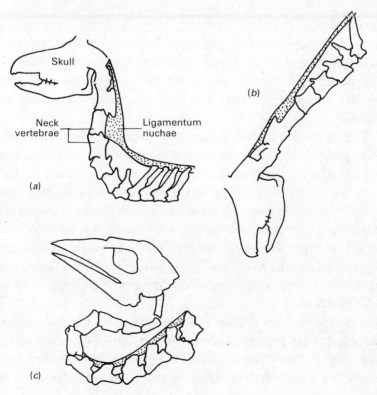

fresh carcases of sheep and deer to expose the ligament, while disturbing the surrounding tissues as little as possible. We stuck pins into the ligament at intervals along its length, so that the same points in the ligament could be located when the head and neck were placed in various positions. We took X-ray pictures of the neck in positions simulating a normal alert posture (Fig. 2.2(a)) and the feeding posture (Fig. 2.2(b)). The pins were visible in these pictures and we measured the distances between them, correcting for the magnification that inevitably occurred because the plate was further than the specimen from the X-ray source. Next, we dissected out the ligament, which shortened because it had been under tension even in the alert position of the intact neck. We measured the distances between the pins again and found that the strain in the intact sheep had been about 0.3 in the alert position and 0.8 in the feeding position. (These are approximate means of strains that varied along the ligament.) We then investigated the properties of the ligament as already described (section 1.2) and were able to show that the tension in the ligament must have been about 10 N in the simulated alert position (Fig. 2.2(a)) and 20–80 N in the feeding position (Fig. 2.2(b)). The largest forces occurred in the thick part of the ligament, at the proximal end of the neck. Further calculation showed that the forces in the feeding position would support most of the weight of the head and neck, so that little tension would be needed in neck muscles. Similar experiments with deer and a camel gave broadly similar results.

The ligamentum nuchae gives very little support to the neck in the alert position, but the muscle forces required then are quite small because the centre of mass of the head and neck is less far forward (relative to the trunk) than in the feeding position. It reduces the tension needed in the muscles in the feeding position, saving some of the metabolic energy that would otherwise be needed to maintain muscle tension. The work that has to be done to stretch it is not done by muscles, but by gravity. (This is a difference from the inner hinge ligament of bivalve molluscs, which has to be compressed by the adductor muscle.).

The ligamentum nuchae is arranged so that natural neck movements produce strains of about 0.8. It cannot be stretched much more without breaking, so most of the available range of extension is used. The same is true of elastic ligaments in birds that have similar properties and serve the same function, although they are differently arranged (Bennett & Alexander, 1987). They are short ligaments, connecting each vertebra to the next (Fig. 2.2(*c*)). They lie very close above the vertebral centra and would suffer rather small strains, as the head was raised and lowered, if they formed a continuous strand. In fact, they are short enough for the strains in them to reach 0.6, when a turkey (*Meleagris gallopavo*) feeds from the ground. It seems that neck ligaments, like the hinge ligaments of scallops (section 2.1), tend to be arranged so as to use most of the available range of strain.

2.3 Sea anemones

The mesogloea of sea anemones is another material that serves as a muscle antagonist, or rather as an antagonist both for muscles and for the cilia that can inflate the animal by pumping in water. Unlike the ligaments of mollusc shells and vertebrate necks, it is not associated with any rigid skeleton.

Fig. 2.3(*a*) is a grossly simplified diagram of a sea anemone. It is a hollow cylinder with a mouth surrounded by tentacles at the top. The body wall is mesogloea (section 1.4) covered inside and out by a single

Fig. 2.3.(*a*) A diagrammatic vertical section through a sea anemone. Cells are shown as rectangles and mesogloea is shown black. (*b*) Drawings traced from photographs of the same individual sea anemone (*Metridium*) with its body inflated to different volumes. The scale is the same in every case. From R. B. Clark (1964). *Dynamics in Metazoan Evolution*. Oxford: Oxford University Press.

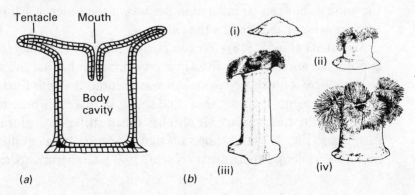

Tentacle   Mouth

Body cavity

(i)

(ii)

(iii)

(iv)

(*a*)   (*b*)

layer of cells. There are incomplete partitions (septa) in the body cavity, but these are not illustrated.

Some species can make spectacular changes of size and shape. *Metridium*, shown in Fig. 2.3(b)), can contract to a small conical mass (i) or inflate to a tall column (iii) or a stout one (iv). It can contract in a few seconds when danger threatens but needs about an hour to re-inflate. Contraction is powered by muscles, but re-inflation by cilia at the corners of the slit-shaped mouth. These cilia produce only small pressure differences. Batham & Pantin (1950) used a manometer to record pressures inside *Metridium*. Pressures up to 150 Pa above that of the surrounding water occurred briefly and seemed to be produced by muscles, but pressures of around 30 Pa were maintained for long periods and were probably due to the cilia.

The muscles and cilia can be made inactive by putting the animal in a mixture of seawater with an isotonic solution of magnesium chloride, or in seawater with added menthol. Both treatments have the same effect: whether the animal is initially inflated or contracted, it changes gradually to the state shown in Fig. 2.3(b)(ii) (Alexander, 1962). This is presumably the size and shape at which the mesogloea is unstrained: it was shown in section 1.4 that the mesogloea is elastic. Inflation to larger sizes stretches the mesogloea, and contraction to smaller sizes compresses and buckles it.

Mesogloea shows a much higher Young's modulus to forces that act for only a few seconds, than to forces that continue for several hours (section 1.4). This property seems appropriate to the way of life of sea anemones. They live on shores, attached to rocks. They are exposed to waves which exert forces with periods of the order of 10 s. They have only a weak ciliary pump to inflate the body. The mesogloea resists waves reasonably well, maintaining the shape of the animal, but can be inflated in periods of the order of 1 h by the small pressure generated by the cilia.

Metridium (Fig. 2.3(b)) lives in fairly calm water. *Anthopleura* is another sea anemone that lives in a very different habitat, on rocky shores where it is exposed to violent wave action. It avoids the fastest water movements by being short and squat, so that the whole animal is close to the rock surface. It also has much stiffer mesogloea than *Metridium* (Fig. 1.7) and shows no sign of approaching equilibrium even in very long experiments (Koehl, 1977). (Experiments cannot be prolonged indefinitely because decay sets in.) Its stiff mesogloea prevents it from changing its size and shape like *Metridium*, but helps it to withstand wave action. The extra stiffness is partly due to a higher concentration of protein in the matrix, and partly to the presence of more filler (collagen fibres) (see section 1.6).

3

Springs as energy stores: running

This chapter is about running by fairly large mammals, of the size of cats and larger. It will be shown that dogs, horses, kangaroos and people save energy by elastic mechanisms, when they run. Walking uses a quite different energy-saving principle, which does not involve elastic mechanisms and is not discussed here (see Alexander, 1980).

3.1 The bouncing ball principle

Wheeled vehicles can be very economical of fuel, for travel over smooth, level ground. They need power to overcome friction in their moving parts, but this can be small if they are well lubricated. They also need power to overcome air resistance, but this is small at low speeds. The advantage of wheels is illustrated by a comparison of running and cycling at speeds low enough for the person not to accumulate an oxygen debt, so that the rate of oxygen consumption is proportional to the rate of metabolic energy consumption. A cyclist uses oxygen less than half as fast as a runner travelling at the same speed (Pugh, 1974).

Animals have nevertheless not evolved wheels. Those that travel fast over land run on legs. It is difficult to imagine how wheels could evolve from the materials and structures found in animals: how could a wheel be supplied with blood vessels and nerves so that they did not get intolerably twisted as the wheel rotated? Also, it is arguable that, though wheels work well on roads, they are unsatisfactory on rough ground. They cannot step over obstacles in the way legs can (La Barbera, 1983). Legged vehicles have been proposed as the ideal for very rough terrain, and some prototypes have been built (Todd, 1985).

This chapter shows how the extra energy costs, incurred by using legs instead of wheels, are reduced by elastic mechanisms.

First, we must understand how the costs arise. Fig. 3.1(*a*) shows three stages of a step of an animal running from left to right. The leg is represented as having just two joints, a hip and a knee. We will consider what muscle action is needed, in the course of the step. The mass of the leg will be ignored (but the effects of leg mass will be considered later in the chapter).

The forces feet exert on the ground must have vertical components, to support the animal's weight. It is assumed in Fig. 3.1(*a*) that the force is precisely vertical, throughout the step. In all three

Fig. 3.1. Diagrams of a running animal, showing the trunk and one leg. (*a*), (*b*) and (*c*) show three possible patterns of muscle action. (i), (ii) and (iii), in each case, show successive stages of a step.

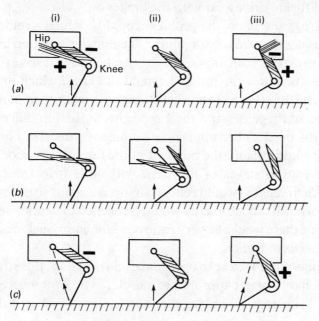

stages it exerts a clockwise moment about the knee, tending to bend the knee. Therefore, a knee extensor muscle must be active throughout the step. At stage (i) the force on the foot exerts an anticlockwise moment about the hip, so a hip extensor muscle must be active; at stage (ii) it passes through the hip, exerting no moments; and at stage (iii) it exerts a clockwise moment about the hip, requiring a hip flexor to be active. Only the muscles that must be active at each stage are shown in the figure.

The hip starts flexed, and extends progressively during the step. The active hip extensor muscle at stage (i) must be shortening, and therefore doing work. This is indicated by a plus (+) sign. The active hip flexor at stage (iii) must be extending. It is not doing work: rather, it is acting as a brake, degrading mechanical energy to heat. This is commonly expressed by saying it does *negative* work, so the muscle is marked with a minus (−) sign.

The knee is bending at stage (i) and extending at stage (iii). The knee extensor muscle is extending, doing negative work, at stage (i) and shortening, doing positive work, at stage (iii). Notice that, both at stage (i) and at stage (iii), one muscle does positive and another negative work. One muscle's work is degraded by the other to heat. The arrangement wastes energy for the same reason that it would be wasteful of fuel to drive a car with the brakes on.

Fig. 3.1(*b*) shows how the problem could be largely avoided, by having muscles that cross both joints. The muscles required are one that tends to extend both joints, which must be active at stage (i), and one that tends to flex the hip and extend the knee, which must be active at stage (iii). (Both are needed at stage (ii), when their moments at the knee add together and their moments at the hip cancel each other.) If the muscles had appropriate moment arms about hip and knee, they could balance the moment of the force on the foot about both joints, at all stages of the step, with little help from other muscles. Their lengths would remain almost constant, so they would do little work. Metabolic energy would be needed to develop tension in them, but there would be very little need for additional metabolic energy to perform work.

Two things make it clear that mammals do not adopt this strategy. First, they have not got appropriate muscles. The femorococcygeus muscle of kangaroos resembles the muscle shown in Fig. 3.1(*b*)(i)

(Alexander & Vernon, 1975), but other mammals lack this muscle. All mammals seem to have a rectus femoris muscle arranged like the muscle of Fig. 3.1(*b*)(iii), but it generally forms only about one-quarter of the knee extensor musculature. It is probably not strong enough to provide by itself the knee moments needed in running, and it has been shown by electromyography in several species that the other knee extensors are also used, (see Rasmussen, Chan & Goslow, 1978). (Electromyography means recording the electrical events in muscle that accompany mechanical activity.)

The second kind of evidence, that the strategy shown in Fig. 3.1(*b*) is not used, comes from force platform records. A force platform is an instrumented panel that can be set into the floor. Forces exerted, for example, by an animal stepping on it, are indicated by electrical outputs. Separate outputs can provide measurements of the vertical, longitudinal and transverse components of the force. Force platform records have been made of running by various mammals including dogs, sheep, kangaroos and people (see, for example, Cavagna, Heglund & Taylor, 1977). They show that the force on a foot does not remain vertical, as in Figs 3.1(*a*) and (*b*). Instead, it slopes, as in Fig. 3.1(*c*). At stage (i) it slopes backward, tending to decelerate the animal, and at stage (iii) it slopes forward, tending to accelerate it. The figure shows the force in line with the hip at all stages, but this is not quite realistic. It is more usually kept in line with a point a little above the hip or shoulder (Jayes & Alexander, 1978).

Fig. 3.1(*c*) is a more realistic model of running, than either (*a*) or (*b*). If the force on the foot is always in line with the hip joint, no tension will be needed in hip muscles. (Remember our assumption that the mass of the leg is negligible. If the leg has appreciable mass, torque will be needed at the hip to balance the moment of the weight of the leg, and to accelerate the leg.) However, tension is needed in a knee extensor muscle throughout the step. This muscle is lengthening, doing negative work, in Fig. 3.1(*c*)(i), and shortening, doing positive work, in (iii).

This strategy has an important advantage. The negative work is followed by positive work *done by the same muscle*. The muscle can be replaced by a spring which will have the same effect, but will not consume metabolic energy. This is the principle of pogo-sticks, and of bouncing balls. A ball bouncing on a rigid surface will make many bounces without any fresh input of energy. A perfectly elastic ball in a frictionless world would continue bouncing for ever.

The horizontal components of force in Fig. 3.1(c) decelerate and re-accelerate the body, making it lose and then regain kinetic energy. In running, there are instants when both feet (of a biped) are simultaneously off the ground, so the animal must rise and fall, gaining and losing potential energy. At stage (i) kinetic energy plus potential energy is being lost and at stage (iii) it is being regained. This energy must be stored as elastic strain energy at stage (ii), if the bouncing ball principle operates.

3.2   Energy fluctuations in human running

The first persuasive evidence that the bouncing ball principle might be important came from an analysis of human running (Cavagna, Saibene & Margaria, 1964). This evidence depended on calculations of kinetic and potential energy fluctuations. We need to understand how these energies are calculated.

A human or animal body is a complex structure whose parts move relative to each other. Think of a body of mass M as a collection of n particles of masses m_1, m_2, m_3, etc. At time t the coordinates of these particles are (x_1, y_1, z_1), (x_2, y_2, z_2), etc. and their velocities have components $(\dot{x}_1, \dot{y}_1, \dot{z}_1)$, etc. The centre of mass is at (X, Y, Z) and its components of velocity are $(\dot{X}, \dot{Y}, \dot{Z})$. The potential energy can be calculated from the height of the centre of mass

$$\text{potential energy} = MgY \tag{3.1}$$

where g is the gravitational acceleration.

The kinetic energy can be thought of conveniently as the sum of two components. The *external* kinetic energy is the KE due to the movement of the centre of mass

$$\text{external kinetic energy} = \tfrac{1}{2}M(\dot{X}^2 + \dot{Y}^2 + \dot{Z}^2) \tag{3.2}$$

The *internal* kinetic energy is due to movements of parts of the body relative to the centre of mass.

$$\begin{aligned}
\text{internal kinetic energy} \\
= \tfrac{1}{2}m_1 \,[(\dot{x}_1 - \dot{X})^2 + (\dot{y}_1 - \dot{Y})^2 + (\dot{z}_1 - \dot{Z})^2] \\
+ \tfrac{1}{2}m_2 \,[(\dot{x}_2 - \dot{X})^2 + (\dot{y}_2 - \dot{Y})^2 + (\dot{z}_2 - \dot{Z})^2] \\
+ \cdots
\end{aligned} \tag{3.3}$$

Cavagna *et al.* (1964) made films and force platform records of people running. They could have measured height changes from the films and used them to calculate potential energy changes, but this would have been difficult because the centre of mass has no fixed position in the body: it moves as the parts of the body move relative to each other. Instead, they calculated potential energy changes indirectly, from the record of the vertical component of force (see Cavagna, 1975). Similarly, they calculated external kinetic energy changes from the force platform record. Only internal kinetic energy was calculated from the film.

They found that a 70 kg man loses and regains about 115 J (kinetic plus potential) energy per metre travelled, when running at any speed between 3 and 6 m s^{-1}. Muscles seem unable to do positive work with efficiencies above about 0.25, and also use metabolic energy when doing negative work. Therefore, if no elastic energy savings are made, men can be expected to use at least 115/0.25 = 460 J metabolic energy, per metre travelled. However, measurements of oxygen consumption indicated that the actual energy requirement was only about 300 J m^{-1}. The discrepancy suggests that the bouncing ball principle operates, saving a lot of energy. A similar analysis of hopping kangaroos suggested that they too saved energy, in the same way (Cavagna *et al.*, 1977).

3.3 Kangaroos

To make the suggestion convincing, we need to show where the strain energy is stored. Evidence about this came from experiments with a wallaby, *Macropus rufogriseus*, a small species of kangaroo (Alexander & Vernon, 1975). We made it hop across a force platform, and filmed it at the same time. The force record and film were used to calculate the forces exerted by major muscle groups, and the length changes of the muscles.

Fig. 3.2(*a*) represents a frame from the film. The crosses had been drawn on the skin over the hip, knee and ankle joints. A synchronizing device enabled us to identify the instant on the force record, corresponding to this frame. This showed a force of 420 N, at 90° to the horizontal. This was the total force on the two hind feet, so half of it is shown acting on the nearer foot in Fig. 3.2(a). It is assumed to act half way along the area in contact with the ground.

Fig. 3.2(b) shows how we did the calculations for one particular muscle, the gastrocnemius. This muscle runs from femur to heel, crossing both knee and ankle. Its moment arms about these joints, r_k and r_a, had been obtained from X-radiographs (see Alexander & Dimery, 1985a). The angles θ_k and θ_a, and distance s, had been measured from Fig. 3.2(a). The force F on the foot exerts a moment Fs about the ankle joint. The weight of the foot and the inertia force corresponding to the acceleration of the foot also exert moments about the ankle, but these are small enough to be ignored. If the gastrocnemius were the only active ankle muscle, the force in it would have to be Fs/r_a. In fact it cooperates with another strong muscle, the plantaris, and an assumption had to be made about the relative sizes of their forces.

The gastrocnemius tends to get longer as θ_a increases and shorter as θ_k increases. If they increase by $\Delta\theta_a$ and $\Delta\theta_k$, the gastrocnemius lengthens by $(r_a \cdot \Delta\theta_a - r_k \cdot \Delta\theta_k)$. Similar calculations were also made for other muscles, which are shown in Fig. 3.3.

The hamstring and adductor muscles are extensors of the hip joint. The calculations showed that they shortened, exerting force, throughout the period when the foot was on the ground. Thus, they did positive work. Some strain energy must have been stored in their

Fig. 3.2.(a) An outline traced from a film of a wallaby hopping across a force platform, showing the force acting on one foot. (b) A diagram showing how the force in the gastrocnemius muscle, and its changes of length, were calculated.

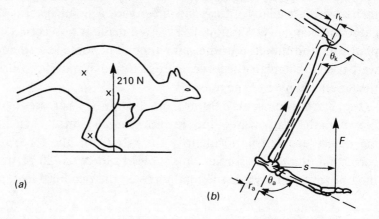

(a)

(b)

muscle fibres and in their (very short) tendons as they developed tension, but that saved no energy. It simply meant that the muscles had to do more work as tension increased (to store the strain energy) and less work later (during the elastic recoil).

The quadriceps muscles are the extensors of the knee. The calculations showed that they exerted force and lengthened, throughout the time that the foot was on the ground. Thus, they did negative work, and again no energy was saved.

The gastrocnemius and plantaris muscles are the principal extensor muscles of the ankle. (In people, the plantaris is rudimentary and the soleus, another ankle extensor, is large.) The calculations showed that both lengthened while the forces in them were increasing and shortened while the forces were decreasing. Thus, they lengthened and shortened like springs, doing negative followed by positive work.

The tendon of the plantaris muscle goes round the heel, right down to the toes, so it is a toe flexor as well as an ankle extensor. The deep digital flexor also goes round the ankle but has a very small moment arm there and is principally a flexor of the toes. It too seems to lengthen and then shorten while exerting force.

Thus, the gastrocnemius, plantaris and deep flexor muscles all seem to be possible strain energy stores, that may save energy in hopping. All of them have short muscle fibres (about 20 mm, in a 17 kg wallaby) and long tendons (280–470 mm) (Ker *et al.*, 1986).

Fig. 3.3. The skeleton, and some of the principal muscles, of a hind leg of a typical mammal. The soleus muscle, which is large in humans but small or absent in many other mammals, is omitted. From R. McN. Alexander (1984). *Am. Zool.*, **24**, 85–94.

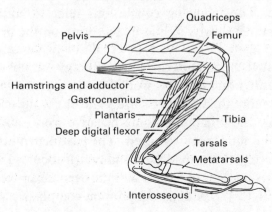

Section 1.7 shows that much more strain energy must therefore be stored in the tendons than in the muscle fibres. Morgan, Proske & Warren (1978) confirmed this for the gastrocnemius by direct experiment. The tendons of these distal leg muscles are the important springs in the legs of kangaroos.

How much energy they save must depend on their compliance. If they had the optimum compliance, they would stretch enough to allow the joint movements without any length change of the muscle fibres (apart from slight elastic stretching). The contractile apparatus of the muscle would have to develop tension, but would do no work. Metabolic energy would be needed to develop tension, but the extra energy needed to perform work would be saved. If the compliance were not optimal, the muscles would have to do some work. If it were less than optimal, the muscle fibres would have to do some of the negative work, and then some of the positive work. If it were more than optimal, the muscle fibres would have to shorten while the tendon was stretching (to compensate for excess stretching) and lengthen while the tendon was shortening: they would have to do positive followed by negative work.

Ker *et al.* (1986) tried to discover whether the compliances of the tendons were optimal. We made tensile tests on the tendons in question, in a dynamic testing machine (see section 1.3). These told us how much strain energy was stored, for any given force on the tendon. We combined this information with tendon forces calculated from Alexander & Vernon's (1975) force platform data, to calculate the strain energy stored in hopping. The animals available for the tendon tests were unfortunately larger (18 kg) than the one used for the force platform observations (11 kg), so allowance had to be made for the difference. The following conclusions refer to an 18 kg animal, hopping rather slowly. While its feet were on the ground it would lose and regain 39 J. This is external kinetic energy plus gravitational potential energy: internal kinetic energy was not calculated. The total positive and negative work done by the muscles and tendons would be more than this, because some of the muscles do positive work while the others are doing negative work, and vice versa, increasing the total work required. The gastrocnemius tendons of the two legs together would store and return about 6 J strain energy, the plantaris tendons about 8 J and the deep flexor tendons about 1 J, a total of 15 J. Tendons of optimum compliance would have stored the whole of the 39 J, but 15 J seems quite a substantial saving. The tendons were stiffer than optimal for the low speed investigated, but may have had optimal compliance for higher speeds, as will be shown in the next section.

Kangaroo rats (*Dipodomys*) are very much smaller than kangaroos, with masses around 100 g. They have long hind legs like kangaroos, and hop like kangaroos. Biewener, Alexander & Heglund (1981) looked for elastic energy storage in their hopping. We used X-ray instead of light cinematography, but our methods were otherwise like those of Alexander & Vernon (1975). We found that strain energy accounted for a much smaller fraction of the work than in kangaroos, largely because the tendons were thicker in proportion to the forces they had to transmit. No similar investigations have been made on other small species, but it seems possible that elastic storage is important in running only in the larger mammals. It has also been argued that it is important in ostriches (*Struthio camelus*, about 40 kg) (Alexander, Maloiy, Njau & Jayes, 1979), but not in quail (*Coturnix coturnix*, about 100 g) (Clark & Alexander, 1975).

3.4 Mathematical models

Fig. 3.4 shows a very simple model of kangaroo hopping. The animal moves in a system of Cartesian coordinates, with the

Fig. 3.4. A mathematical model of a kangaroo hopping.

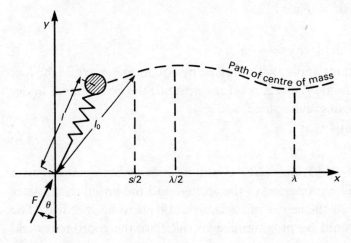

x-axis at ground level. It has a body of mass m and a single leg of negligible mass. This leg is a compression spring of stiffness S, joined to the body by a frictionless joint at the centre of mass. It has a small foot at its end. The leg has length l_0 when there is no force on the foot, and length l when compressed by a force F.

$$F = S(l_0 - l) \tag{3.4}$$

This force acts along the line joining the foot to the centre of mass, at an angle θ to the vertical.

The animal makes strides of length λ, setting down the foot at points $(0, 0)$, $(\lambda, 0)$, $(2\lambda, 0)$, etc. At time t the centre of mass is at (x, y). Consider the half stride in which x increases from 0 to $\lambda/2$. Initially, the foot is on the ground, the leg is vertical, the spring is compressed and the centre of mass is moving horizontally. A little later, the leg is sloping, the spring is extending and the centre of mass is rising. Eventually, the leg reaches its unstrained length l_0, at which time the foot leaves the ground. The step length s is the distance travelled while the foot is on the ground, so it leaves the ground when $x = s/2$. When the half stride ends, the centre of mass is again travelling horizontally, but with the foot still off the ground.

While the foot is on the ground, the horizontal and vertical components of the force on it are

$$F \sin \theta = Fx/l$$

and

$$F \cos \theta = Fy/l.$$

Thus, the equations of motion of the body are

$$\mathrm{d}^2x/\mathrm{d}t^2 = Fx/ml$$

and

$$\mathrm{d}^2y/\mathrm{d}t^2 = Fy/ml - g \tag{3.5}$$

These equations apply, with F given by equation 3.4, while the foot is on the ground. While it is off the ground, the body moves under the influence of gravity alone, and

$$\mathrm{d}^2x/\mathrm{d}t^2 = 0$$
$$\mathrm{d}^2y/\mathrm{d}t^2 = -g \tag{3.6}$$

Equations (3.5) and (3.6) could be solved by iteration on a computer. The properties of the spring, and the initial coordinates and velocity of the centre of mass, would have to be specified. The computer would be programmed to calculate the coordinates and velocity after successive short increments of time. The value that x was found to have when $l = l_0$ would be the half step length, $s/2$. Its value when the vertical component of velocity returned to zero would be the half stride length, $\lambda/2$. The mean speed could be calculated from the duration of the half stride. Thus, step length, stride length and mean speed could be calculated, for given initial conditions.

Three points about this model should be noticed. First, it is grossly simplified. We have specified a massless leg, a frictionless joint and (by implication) a Hookean spring and negligible air resistance. These simplifications may seem reasonable (and the model would be cumbersome without them) but we should remember that they are there. Secondly, the model does not do exactly what we might want. We might want to specify step length, stride length and speed, and calculate the spring stiffness required. Instead, we have to specify spring stiffness and initial conditions, and calculate the other quantities. We can discover what we want to know only by trial and error. Thirdly, the equations cannot be solved analytically, but only numerically. We cannot get an equation giving spring stiffness in terms of step length, stride length and speed.

Alexander & Vernon (1975) avoided the latter two awkwardnesses by abandoning the assumption of a Hookean spring. Instead, we assumed that the vertical component of the force on the foot rose and fell like a half cycle of a sine wave. This may seem arbitrary, but it imitated force platform records quite well. Calculations using that version of the model show that optimum spring stiffness increases, as hopping speed increases. No tendon can have optimum stiffness for hopping at all speeds.

That calculation used particular values of s and λ, chosen to imitate the hopping of real kangaroos at chosen speeds. A more general argument will show that optimum stiffness almost inevitably increases, as speed increases.

Kangaroos hopping (and other mammals running) increase speed partly by increasing stride frequency, but mainly by taking longer strides. Consequently, they cannot keep their feet on the ground for as long a fraction of the stride, at higher speeds. This means that, while the feet are on the ground, they must exert larger forces. (The mean vertical force, averaged over a complete stride, must equal body weight.) The ranges of angular movement of the joints are not very different at different speeds, so the length changes of the muscles remain about the same at all speeds. (Here 'muscles' is intended to include the length of the tendons.) The quantities of work done by each muscle are therefore about proportional to the force it exerts. However, the strain energy stored in each tendon is about proportional to (force)^2 (equation 1.5). Therefore, if tendon stiffness is constant, strain energy accounts for increasing fractions of the work done, as speed and force increase. The optimum stiffness is higher for high speeds, than for low ones.

3.5   The arch of the foot

The initial evidence that the bouncing ball principle might be important in running came from observations on people (section 3.2) but kangaroos were used to demonstrate the importance of tendons (section 3.3). Now we return to people for evidence of a spring in the arch of the foot (Ker *et al.*, 1987).

Fig. 3.5(*a*) shows forces on the foot of a 70 kg man running at a moderate speed. It is based on force records by Cavanagh & Lafor-

Fig. 3.5.(*a*) A diagram showing forces on the foot of a 70 kg man running at 4.5 m s^{-1}. (*b*) An experiment in which similar forces were applied to an amputated foot, in a dynamic testing machine. From R. F. Ker *et al.* (1987). *Nature*, **325**, 147–9.

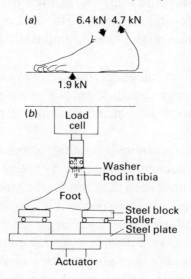

tune (1980) who used a force platform that indicated the point of application of the force on it, as well as the size and direction. It shows the instant when this force is at its peak of about 1.9 kN (2.8 times body weight). This force acts on the ball of the foot (the heel is rising off the ground) and exerts a large moment about the ankle. To balance this, a calculated force of 4.7 kN is needed in the Achilles tendon (the tendon of the gastrocnemius and soleus muscles). The third force, 6.4 kN, is the reaction at the ankle joint that is needed for equilibrium. Together these three forces tend to bend the foot, flattening the arch. Films of barefoot runners show that the arch is perceptibly flattened, at this stage of the step.

Fig. 3.5(*b*) shows an experiment designed to simulate this pattern of forces. The specimen is a foot that had to be amputated because of vascular disease. It is mounted in a dynamic testing machine in which it is compressed. Two steel blocks press upwards on the ball of the foot and on the heel, and the load cell presses down on the cut end of the tibia. Thus, three forces act on the foot, simulating the three forces shown in Fig. 3.5(*a*). Notice that the upward pull of the Achilles tendon is simulated by an upward *push* on the calcaneus (heel bone). (The soft tissue has been removed from the heel so that the steel block presses directly on the bone.) The rollers under the steel blocks allow the foot to lengthen as the arch flattens.

Fig. 3.6(*a*) shows how squeezing in the machine flattened the arch of the foot. The distortion of the foot that occurs in running is similar. Fig. 3.6(*b*) shows the record of one of the experiments in the machine. The actuator was rising and falling, applying force and removing it, at a frequency chosen to simulate running. The graph of force against displacement forms a fairly narrow loop, showing that strain energy was being stored in the foot and largely returned in an elastic recoil: the foot behaved like a spring.

The force registered by the load cell (and shown in Fig. 3.5(*b*)) was the force on the tibia. To simulate running, it should reach 6.4 kN (Fig. 3.5(*a*)). Unfortunately, we could not apply such large forces because the calcaneus was crushed where the steel plate pressed on it. However, we applied forces up to 4.7 kN and were able to extrapolate our measurements, to estimate the strain energy that would be stored when the force was 6.4 kN. We also used the dimensions of the Achilles tendon to calculate the strain energy that would be stored in

Fig. 3.6.(*a*) Outlines of X-ray pictures of a foot, taken during the experiment shown in Fig. 3.5(*b*). Continuous lines show outlines of bones in the unloaded foot, and broken lines show them under a load of 4 kN. (*b*) A record from the experiment shown in Fig. 3.5.(*b*). The force registered by the load cell is plotted against the vertical displacement of the actuator. From the data of Ker *et al.* (1987).

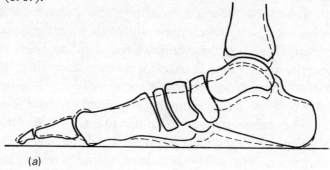

(*a*)

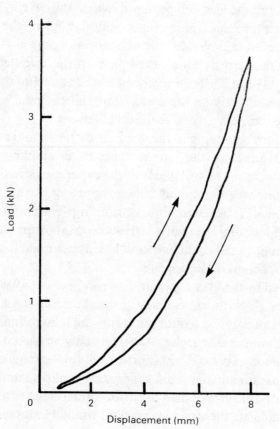

(*b*)

it, when stretched by a force of 4.7 kN (Fig. 3.5(*a*)). Here are our conclusions. Remember that they have been obtained from crude experiments on diseased feet, and may not reflect very accurately the energy stored in healthy feet of living runners.

Each time he sets down a foot, a 70 kg man running at 4.5 m s^{-1} loses and regains about 100 J (external kinetic plus potential) energy. Of this, we estimate that about 35 J is stored as strain energy in the Achilles tendon, and 17 J as strain energy in the arch of the foot. Thus, substantial savings of energy are made, and both the tendon and the arch are important.

In some of our experiments we cut the ligaments of the arch in turn and found that less strain energy was stored, for the same movement of the machine, after each cut. Many different ligaments are apparently important, in the energy-storing mechanism. The most prominent of these is the plantar aponeurosis, which runs along the sole from the heel to the ball of the foot, but it seems to be no more important than some other ligaments that lie deeper among the bones.

The arch of the foot is a human peculiarity, but its joints have homologues in other mammals. Distortion of the hind feet, similar to the flattening of the human arch, has been demonstrated in running camels (Alexander *et al.*, 1982).

3.6   Hoofed mammals

The elastic properties of the tendons of kangaroos and people reduce the work their muscles have to do in hopping or running. However, energy is still needed to develop tension in the muscles. Short muscle fibres can exert as much force as an equal number of long ones, but presumably need less energy to activate them. It may therefore be possible to make running more economical, by evolving shorter muscle fibres. Some large muscles with long fibres will be needed to do the positive and negative work needed for rapid acceleration and deceleration, and for jumping, but there may be an advantage in shortening the fibres of distal leg muscles, whose tendons serve as springs in running.

This has happened in the evolution of ungulate mammals, as Table 3.1 shows. The wallaby is the only biped in the table and the horse is

Table 3.1. *Lengths in millimetres of the muscle fibres in hind leg muscles of some mammals. The fibres had these lengths in dissections of animals fixed with their legs extended, but would have lengthened and shortened a little as the living animal moved. From data collected by the author and his colleagues*

		Lengths of muscle fibres (mm)				
Mammal	Body mass (kg)	Gastrocnemius	Soleus	Plantaris	Deep digital flexor	Interosseus
Wallaby	17	21	28	18	19	7
Baboon	15	32	34	32	22	**
Dog	24	25	*	10	20	**
Gazelle	29	11	*	2.5	16	**
Horse	approx. 250	48	*	1–2	30	<1

* Rudimentary or absent. ** No data.

much larger than the rest, but the other animals are quadrupeds of fairly similar size. The ungulates (gazelle and horse) have remarkably short fibres in some of their muscles, especially in the plantaris and interosseous and in similarly placed muscles in the foreleg. In the plantaris of the camel (*Camelus dromedarius*) it is difficult to find any muscle fibres at all: the tendon runs almost uninterrupted from the origin on the femur to the insertion on the toes (Alexander *et al.*, 1982).

The rudimentary state of the fibres in some of these animals is illustrated by a simple calculation. The interosseous muscle of the horse has fibres less than 1 mm long, which presumably cannot shorten more than 0.5 mm. It has a moment arm about the metatarsophalangeal joint of 14 mm. A contraction of 0.5 mm would move the joint through only 0.5/14 radians, or 2°. It seems inconceivable that so slight an ability to shorten can have much value. The muscle fibres are mere vestiges preserved with the tendons, which serve as passive springs.

I explained in section 3.3 how muscle length changes in kangaroos were calculated from leg movements seen in films. These were changes in length of the muscle plus its tendon. Dimery, Alexander & Ker (1986) applied a more sophisticated version of the same method to horses. Here, some of the muscles had rudimentary fibres and their length changes must have been almost entirely due to stretching or slackening of the tendons. We found that such tendons stretch 2–6% in walking, 3–7% in trotting and 4–9% in galloping. Some of them become very slack at other stages of the stride.

Lochner, Milne, Mills & Groom (1980) had previously made more direct measurements of tendon strain in horses. They performed surgical operations on horses to attach strain gauges to leg tendons, so that they could record the strains involved in locomotion after the horses had recovered. Their results show peak strains of 6% or more in several tendons of the foreleg, in brisk walking.

My colleagues and I have investigated strain energy storage in the tendons of donkeys (*Equus asinus*) and two species of deer (Alexander & Dimery, 1985b; Dimery & Alexander, 1985; Dimery, Ker & Alexander, 1986). Our methods were like those used for wallabies (section 3.3) supplemented by experiments with feet in the dynamic testing machine, much like the experiments on human feet (Fig. 3.5). Unfortunately, the forces involved in running had to be estimated, because we had no force platform records. Our conclusions were that both donkeys and deer make very large savings of energy by elastic storage, when they run fast. Most of the negative and positive work done by each leg is accounted for by strain energy storage in tendons.

3.7 Galloping

Most quadrupedal mammals use two kinds of running. At moderate speeds they trot, but at higher speeds they gallop. Fig. 3.7 shows the difference. In trotting, diagonally opposite feet move together, the left forefoot with the right hind and the right fore with the left hind. In galloping the two forefeet are set down (not quite simultaneously) and then the two hind feet. Hoyt & Taylor (1981) trained small ponies to run on a treadmill, using different gaits on command. They measured the ponies' rates of oxygen consumption and found that trotting consumed less oxygen than galloping at speeds below 4.5 m s^{-1}. However, at speeds above 4.5 m s^{-1}, galloping used less oxygen than trotting. When the ponies were allowed to move at will in a paddock, they trotted at 2.8–3.8 m s^{-1} but galloped at speeds over 5 m s^{-1}. They chose whichever gait was more economical of oxygen (and so of energy) for the speed at which they were moving.

Alexander, Dimery & Ker (1985) made a suggestion that seems to explain why galloping is more economical than trotting at high speeds. The elastic mechanisms described so far in this chapter exchange external kinetic and potential energy with elastic strain energy. They act both in trotting and in galloping, but the additional mechanism to be described now exchanges internal kinetic energy with strain energy, and acts only in galloping.

Internal kinetic energy (equation 3.3) is the energy associated with movements of parts of the body relative to the centre of mass. Large amounts of it are associated with the leg movements of running. Whenever a leg is halted at the end of a forward or backward swing, and made to swing the other way, internal kinetic energy is lost and regained. When an animal runs at speed v, a foot on the ground has speed $-v$ relative to the centre of mass. If the mass of the foot is m, the internal kinetic energy associated with its movement is $\frac{1}{2}mv^2$. Points higher on the leg have smaller velocities relative to the centre of mass, but the internal kinetic energy associated with the movement

Fig. 3.7. Successive stages of the stride of horses (*a*) trotting and (*b*) galloping. From P. P. Gambaryan (1974). *How Mammals Run.* New York: Wiley.

(a)

(b)

of the whole leg is about proportional to v^2. This energy is lost and regained once in each stride. The internal kinetic energy associated with the forward swing of the leg may not increase so fast with increasing speed, but the total fluctuations of internal kinetic energy increase very rapidly. In contrast, force platform measurements show that the external kinetic energy lost and regained in each stride increases only a little with speed, about in proportion to the increased length of the stride, and potential energy fluctuations actually decrease with increasing speed. At low speeds, the fluctuations of external kinetic and potential energy are larger than the fluctuations of internal kinetic energy, but at high speeds the reverse is true (Heglund, Fedak, Taylor & Cavagna, 1982). As speed increases, it becomes increasingly desirable to save as much as possible of the energy needed for internal kinetic energy fluctuations.

The back is bent at one stage of a galloping stride, and extended at another (Fig. 3.7(*b*)). In some mammals (for example, greyhounds) the whole lumbar region bends into a fairly smooth curve but in others (horses) most of the bending is at the joint where the last lumbar vertebra meets the sacrum. Alexander *et al.* (1985) suggested that the back might incorporate a spring that could store and return internal kinetic energy.

The idea is illustrated by Fig. 3.8. The diagrams at the top represent successive stages of a galloping stride. Below, graphs show energy fluctuations. In stage (*a*) the forelegs are doing negative work,

Fig. 3.8. Diagrams of successive stages in a galloping stride, with schematic graphs showing energy changes. 'External energy' means external kinetic plus potential energy. From R. McN. Alexander *et al.* (1985). *J. Zool, A*, **207**, 467–82.

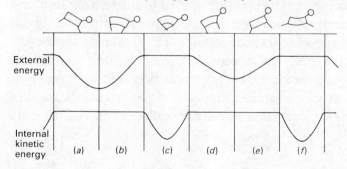

removing (external kinetic plus potential) energy from the body. In (*b*) they are doing positive work, restoring this energy. The hind legs do the same in stages (*d*) and (*e*). These are the energy fluctuations that are largely balanced by storage of strain energy in leg tendons. In stage (*c*) the forelegs reach the end of their backward swing, are halted and start swinging forward. Also, the hind legs reach the end of their forward swing, are halted and start swinging back. Internal kinetic energy therefore passes through a minimum. This is when the back is most bent. A spring that stretched as the back bent could store and then return the internal kinetic energy. Another minimum of internal kinetic energy occurs at stage (*f*), when the back is hyperextended. A spring that resisted hyperextension of the back could store internal kinetic energy then.

The second hypothetical spring has not yet been identified but the first one (that resists bending) has. Alexander *et al.* (1985) pointed out that the longissmus muscle, the main extensor of the back, has a very large aponeurosis. (An aponeurosis is a tendon formed as a broad sheet rather than a cord.) We dissected the aponeuroses from a deer and a dog, performed tensile tests on them, and showed that they had elastic properties like other tendons. When the muscle is taut, it stretches the aponeurosis, storing elastic strain energy. It also compresses the vertebral column. We made compression tests on vertebrae and intervertebral discs and confirmed that they could serve as compression springs. Rough calculations indicated that about half the internal kinetic energy, lost and regained at stage (*c*) of the stride of a deer galloping fast, could be stored as strain energy in the aponeurosis and vertebral column. Some more may be stored in the ligaments that limit the movements of the leg joints.

Such a mechanism could not work in trotting because it depends on the back being bent by opposite inertial torques on the fore- and hind legs. In trotting, each foreleg moves in phase with a hind leg and the necessary opposing torques do not occur. Galloping is probably not advantageous at low speeds because the back movements involved increase the kinetic energy fluctuations a little. It is only at high speeds, at which internal kinetic energy fluctuations are very large, that the savings made possible by galloping exceed the extra costs.

4

Springs as energy stores: swimming and flight

4.1 Principles

Many animals fly by flapping wings or swim by beating a tail. Internal kinetic energy falls and rises as the wing or tail is decelerated and accelerated again, at the end of one stroke and the beginning of the next. It may be possible to save energy as it is also saved in galloping (section 3.7), by having springs that exchange strain energy with internal kinetic energy. The principles have been discussed in the contexts of insect flight (Weis-Fogh, 1972) and dolphin swimming (Bennett, Ker & Alexander, 1987).

In each stroke of an insect's wing or a dolphin's fluke, work is done against the drag of the air or water. This will be described as hydrodynamic work. Also, positive work is done accelerating the wing or fluke at the beginning of the stroke and negative work decelerating it at the end. This will be described as inertial work. It is needed for accelerating not only the mass of the wing or fluke, but also the 'added mass' of fluid that moves with it.

Consider a plate making sinusoidal oscillations in a fluid. It will serve as a model both of the wing movements of insects and of the fluke movements of dolphins. At time t the plate is at position x.

$$x = a \cos \omega t \qquad (4.1)$$

where a is the amplitude of its movements and ω is 2π times the frequency. Differentiation gives the velocity $\dot{x}$ and the acceleration $\ddot{x}$.

$$\dot{x} = -a\omega \sin \omega t \qquad (4.2)$$
$$\ddot{x} = -a\omega^2 \cos \omega t \qquad (4.3)$$

A force F_h is needed to overcome drag and do *hydrodynamic* work. It should be proportional to $\dot{x}^2$ if, as implied by the simple model, the plate makes a constant angle to its path through the fluid. It is negative in the downstroke and positive in the upstroke.

$$
\begin{aligned}
F_h &= \mp K\dot{x}^2 \\
&= \mp Ka^2\omega^2 \sin^2\omega t \quad \text{(using equation 4.2)} \qquad (4.4a) \\
&= \mp Ka^2\omega^2 (1 - \cos^2\omega t) \\
&= \mp K^2\omega^2(a^2 - x^2) \quad \text{(using equation 4.1)} \qquad (4.4b)
\end{aligned}
$$

In addition, a force F_i is needed to accelerate and decelerate the plate, doing *inertial* work. It is the acceleration $\ddot{x}$ multiplied by a mass m that includes both the mass of the plate and the added mass of fluid moving with it.

$$
\begin{aligned}
F_i &= m\ddot{x} \\
&= -ma\omega^2 \cos \omega t \quad \text{(using equation 4.3)} \qquad (4.5a) \\
&= -m\omega^2 x \quad \text{(using equation 4.1)} \qquad (4.5b)
\end{aligned}
$$

Fig. 4.1 shows these forces plotted against position x during an upstroke (a) and a downstroke (b). Areas between the curves and the horizontal (x) axis represent work done on the plate. This is positive work (hatched) if the arrows indicating the direction of motion go clockwise around the area, and negative (stippled) if they go anti-clockwise. The hydrodynamic work is always positive but the inertial work is positive in the first half of each stroke (while the plate is being accelerated) and negative in the second half. Let the quantities of work done in each stroke be W_h (positive) hydrodynamic work, W_i positive inertial work and $-W_i$ negative inertial work.

It might be thought that the total work needed for the stroke would be $(W_h + W_i)$ (positive) and $-W_i$ (negative). This is not true because positive hydrodynamic work and negative inertial work are done simultaneously in the second half and partially cancel each other: kinetic energy removed from the plate is used to do some of the hydrodynamic work. The quantities of work actually required are represented by the areas in the bottom graphs in Fig. 4.1(a) and (b). These graphs show the total force $(F_h + F_i)$. Notice that this changes sign near the end of each stroke, when reverse force is needed to decelerate the plate.

Now consider the power needed to drive the plate. The hydrodynamic power P_h is the rate of doing hydrodynamic work and the inertial power P_i is the rate of doing inertial work. Each is the appropriate force, multiplied by the velocity. The hydrodynamic power is always positive

$$P_h = F_h \dot{x} = |Ka^3\omega^3 \sin^3 \omega t| \tag{4.6}$$

(using equations 4.2 and 4.4a. The vertical lines $|\ \ |$ mean that it is positive, whether $\sin \omega t$ is positive or negative). The inertial power is positive in the first half of each stroke and negative in the second half.

$$\begin{aligned} P_i &= F_i \dot{x} = ma^2\omega^3 \sin \omega t \cos \omega t \\ &= \tfrac{1}{2}ma^2\omega^3 \sin 2\omega t \end{aligned} \tag{4.7}$$

(using equations 4.2 and 4.5a and remembering that for any angle α $\sin 2\alpha = 2 \sin \alpha \cos \alpha$). Finally, the total power needed to drive the plate is the sum of the two components

$$P_{tot} = P_h + P_i = a^2\omega^3 \{|Ka \sin^3 \omega t| - \tfrac{1}{2}m \sin 2\omega t\} \tag{4.8}$$

These powers are shown plotted against time in Fig. 4.1(c). The areas under the graphs represent the same work as the corresponding areas in Fig. 4.1(a) and (b), presenting the same picture in a different way. (Areas under graphs of force against displacement, and of power against time, both represent work.)

Fig. 4.1. Graphs showing the force and power required to drive the oscillating plate, which is described in the text. In (a) the components of force (F_h and F_i) and their total are plotted against position (x) during an upstroke. In (b) the same is done for a downstroke. In (c) the components of power (P_h and P_i) are plotted against time. Hatched areas represent positive work and stippled areas represent negative work.

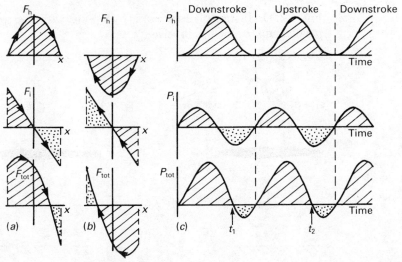

The wing or fluke that is represented by the oscillating plate must be driven by muscles, but the work that the muscles have to do can be reduced by suitably arranged springs. Two possible arrangements are shown in Fig. 4.2. In (*a*) the spring is in parallel with the muscles, so that the force on the plate is the sum of the spring force and the muscle force. This is the arrangement implied by Weis-Fogh (1972) in his analysis of hovering flight.

The spring shown in Fig. 4.2(*a*) exerts a downward force when the plate is above its equilibrium position and an upward force when it is below the equilibrium position. In either case the force is $-Sx$, where x is the displacement from the equilibrium position and S is the stiffness of the spring. The work that the muscles have to do is minimized if the stiffness is given the optimal value S_{opt} which makes the spring force supply the whole of the inertial force

$$-S_{opt}x = F_i = -m\omega^2 x$$
$$S_{opt} = m\omega^2 \tag{4.9}$$

(using equation 4.5*b*). The spring then does all the (positive and negative) inertial work, leaving the muscles to do only the hydro-dynamic work. Thus, the depressor muscle must be active through-out the downstroke, and the elevator muscle throughout the upstroke. Each does work W_h during its stroke.

Notice that the optimum spring stiffness, given by equation 4.9, depends on the frequency ($\omega/2\pi$). A higher frequency requires a stiffer spring.

Fig. 4.2. Oscillating plates powered by muscles with (*a*) a spring in parallel and (*b*) springs in series.

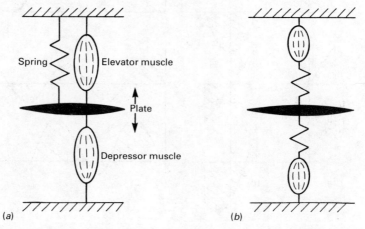

(*a*)   (*b*)

Alternatively, springs may be arranged in series with the muscles, as shown in Fig. 4.2(*b*). This arrangement was analysed by Bennett *et al.* (1987). The force on the plate $(F_h + F_i)$ is not divided between the spring and the active muscle, as in the parallel case, but acts on both. The depressor muscle must start exerting tension before the end of the upstroke, to decelerate the plate, and the elevator muscle must become active before the end of the downstroke. The elevator must be active from time t_1 to time t_2 (Fig. 4.1(*c*)) and the depressor from time t_2.

Let each of the springs in Fig. 4.2(*b*) have stiffness S'. At time t force $(F_h + F_i)$ acts on one of these springs, storing strain energy E,

$$E = \tfrac{1}{2}(F_h + F_i)^2/S' \quad \text{(using equation 1.5)} \qquad (4.10)$$

When the strain energy is increasing, a spring is doing negative work, and when it is decreasing, a spring is doing positive work. The rest of the work has to be done by the muscles. Thus, the power required of the muscles (P_{mus}) is the total power (P_{tot}) plus the rate of change of strain energy $(\dot{E})$.

$$P_{mus} = P_{tot} + \dot{E} \qquad (4.11)$$

where P_{tot} is given by equation 4.8 and $\dot{E}$ can be obtained by differentiating equation 4.10 after putting in values for the forces taken from equations 4.4*a* and 4.5*a*.

Fig. 4.3 shows graphs of P_{mus} against time for systems with parallel

Fig. 4.3. Graphs showing the power required of the muscles $(P_{mus}$, equation 4.11) plotted against time for the model with springs in series (Fig. 4.2(*b*)). In (*a*) the springs have infinite stiffness, in (*b*) they have optimum stiffness (equation 4.12) and in (*c*) they have lower stiffness. Hatched areas represent positive work and stippled areas represent negative work. 'Depressor' and 'elevator' show which muscle is active at each stage of the cycle.

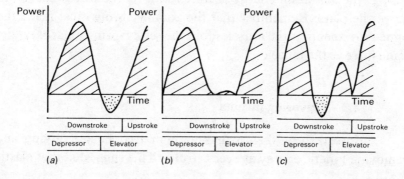

springs of three different stiffnesses. In (a) they have infinite stiffness so the graph is identical with the graph of P_{tot} in Fig. 4.1(c). Each muscle does negative work when it first becomes active, followed by positive work. In (b) the springs have optimum stiffness. Each stretches as the muscle develops tension doing the necessary negative work. It recoils as the muscle tension falls, doing some of the positive work that the muscle would otherwise have to do. The muscle does positive work equal to the hydrodynamic work W_h, and no negative work, in each stroke. Thus, the work required of the muscles is the same as in the parallel system (Fig. 4.2(a)) with optimal stiffness. In Fig. 5.3(c) the stiffness of the springs is too low. Each stretches too much as the muscle develops tension, and the muscle has to compensate by shortening, doing positive work. Each spring shortens too much as muscle tension declines and the muscle has to compensate by lengthening, doing negative work. The muscle has to do more positive work than in (b) and some negative work.

Bennett *et al.* (1987) showed that the optimum spring stiffness for the system with series springs, is

$$S'_{opt} = [1 + (2aK/m)^2]^{1/2} m\omega^2$$
$$= [1 + 0.56(W_h/W_i)^2]^{1/2} m\omega^2 \qquad (4.12)$$

(We expressed the result in terms of a ratio of forces, but the ratio of hydrodynamic to inertial work is preferred here because it can usually be calculated more directly.) Notice that this is larger than the optimum stiffness for a parallel spring (equation 4.9).

We will have to be careful to distinguish, in what follows, between parallel and series elastic elements. The difference in the way they behave is even more striking than the account so far may have shown. The stiffer a parallel spring is, the more strain energy it stores as the plate moves with constant amplitude. However, the stiffer a series spring, the *less* strain energy is stored in it by the forces needed to move the plate. Remember that the parallel spring must make the same movements as the muscles, but the series spring must exert the same force as the muscle.

4.2   Hovering insects

Insects beat their wings at high frequencies, giving and removing kinetic energy at every stroke. This suggested that elastic mechanisms might be capable of saving a lot of energy (Weis-Fogh, 1972).

Hovering is flight in which the animal remains almost stationary, relative to the surrounding air. (The beating of the wings must drive air downwards to support the animal's weight, but the animal is stationary relative to more distant air.) Bees and moths hover to take nectar from flowers, and dragonflies hover while watching for prey. Even when travelling, many insects use styles of flight that resemble hovering, in that the maximum speeds of the tips of the beating wings are much larger than the forward speed of the body.

One type of hovering is common and relatively easy to analyse. Weis-Fogh (1973) called it 'normal hovering'. It is used by bees, wasps, moths, beetles and hummingbirds. The body is tilted at a steep angle and the wings beat horizontally (Fig. 4.4). Each stroke of

Fig. 4.4.(*a*) A moth (*Manduca sexta*) hovering, drawn from a photograph. (*b*) The same, seen from above, traced from a cine film. The undersides of the wings are shown black. From T. Weis-Fogh (1973). *J. exp. Biol.*, **59**, 169–230.

(*a*)

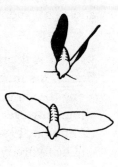

(*b*)

each wing tip describes a little less than a semicircle. The wings have their dorsal sides uppermost (as might be expected) in the forward stroke, but turn over so that the ventral side is uppermost for the backward stroke. Thus, the stiffened anterior edge of the wing is always leading. The wings are tilted at angles of attack that make them drive air downwards in both strokes. More information about the movements involved in normal hovering has been obtained from high-speed films by Ellington (1984a).

Most insects have two pair of wings, but in normal hovering the fore- and hind-wings of each side move as a unit, and it will be convenient to write as if they were united to form a single wing. Consider a wing that beats through an angle $2\gamma_0$ at a frequency $\omega/2\pi$. Its position at time t is given by

$$\gamma = \gamma_0 \cos \omega t \qquad (4.13)$$

and its angular velocity by

$$\dot{\gamma} = -\gamma_0 \omega \sin \omega t \qquad (4.14)$$

so the maximum angular velocity attained in each stroke is $\gamma_0\omega$. The inertial work equals the kinetic energy given to the pair of wings, left and right, and is

$$W_i = I\gamma_0^2\omega^2 \qquad (4.15)$$

where I is the moment of inertia of one wing, left or right, about its base. This moment of inertia depends on the distribution of mass along the wing, and also on the added mass of air that moves with the wing. However, the added mass is relatively small and can be ignored (Ellington, 1984c).

The hydrodynamic (aerodynamic) power needed for hovering has two components, induced power and profile power. The induced power is needed to give kinetic energy to the air that is driven downwards. Simple helicopter theory gives it the value $(m^3g^3/2\varrho A)^{1/2}$, where mg is body weight, ϱ is air density and A is the area swept out by the beating wings. Ellington's (1984b) more realistic vortex theory gives values about 15 % higher. The profile power is the power that would be needed to move the wings through the air even if no lift were being produced. Too little is known about profile drag on oscillating wings for reliable calculations to be possible, but Ellington (1984c) made the best estimates he could, for various insects. The hydrodynamic work W_h can be calculated by multiplying the estimated power by the duration of a wing stroke.

Ellington (1984c) used data taken from his films to calculate the inertial and hydrodynamic power involved in normal hovering, by a few bees and flies and a beetle. The ratio W_h/W_i, calculated from his results, lies between 0.3 and 1.4 in every case. This implies that elastic mechanisms could make substantial savings of energy. They could make much larger proportional savings than suggested by Fig. 4.1, which has been drawn with $W_h/W_i = 2.7$. The larger the inertial forces, and the smaller the ratio W_h/W_i, the bigger the savings that can be made.

Weis-Fogh (1973) used an argument based on muscle efficiency to support his suggestion that elastic mechanisms are important in insect flight. (A similar argument for human running was presented in section 3.2) He chose a few species whose metabolic rates while hovering had previously been measured. He made rough calculations of the hydrodynamic and inertial powers, and compared these with the metabolic powers. The comparison seemed to show that a wasp (*Vespa*), a mosquito (*Aëdes*) and a fly (*Eristalis*) must have elastic energy-saving mechanisms. If they did not, their muscles would have to work with impossibly high efficiencies. Unfortunately, the metabolic rates that he used seem to have been much too low. The insects for which they were measured were attached to apparatus, and their wing movements were probably not supporting all their weight. More recently, several people have managed to measure the oxygen consumption of insects hovering unsupported. Casey (1981) and Ellington (1984c) have repeated Weis-Fogh's calculations, using the new data. They have found no clear evidence of elastic energy-saving mechanisms. Most of Casey's (1981) data for moths are consistent with muscle efficiencies of about 20 %, even in the absence of elastic mechanisms. Ellington's analysis of two bees (*Apis* and *Bombus*) and a fly (*Eristalis*) requires efficiencies of 12–29 % in the absence of elastic mechanisms. We do not know what the efficiency of insect flight muscle is, but these values would be possible for vertebrate muscle.

Thus, the metabolic evidence fails to demonstrate elastic mechanisms. However, simple observations show that the wings of various insects are mounted on elastic structures. The wings of a dead dragonfly (*Aeschna*) rested in horizontal positions. Torques of the order of 1 mN m were needed to raise or lower them to the extreme positions of the wing beat, and when released they returned to the

horizontal (Weis-Fogh, 1972). The stiffness was only about half the
ideal value given by equation 4.9, but the experiment suggested that
elastic mechanisms might give useful energy savings. Similarly, the
wings of locusts rest, when spread, in the down position. A torque is
needed to raise the wings, and they recoil to the down position when
released. Weis-Fogh (1961*a*) measured the torque, and calculated
the strain energy that was stored in the thorax when the wings were
raised. He found that this energy equalled the kinetic energy given to
the wings, in each wing stroke of flight.

Weis-Fogh (1960) discovered a rubber-like protein in various
parts of the thoraxes of insects. He called it resilin, and supposed that
it was important in elastic energy-saving mechanisms. Three structures containing it are shown in Fig. 4.5. Each can be distorted to
large strains (as illustrated) and recoils elastically when released. The
prealar arm (Fig. 4.5(*a*)) and wing hinges (Fig. 4.5(*b*)) of locusts are

Fig. 4.5. Resilin structures from insect thoraxes. The parts shown
white consist mainly or entirely of resilin. (*a*) The prealar arm of a
locust (*Schistocerca*) strained by three different loads. (*b*) The main
hinge of a locust forewing unstrained (above) and strained (below).
(*c*) The elastic tendon of a dragonfly (*Aeschna*) unstrained and
stretched. From T. Weis-Fogh (1960). *J. exp. Biol.*, **37**, 889–907.

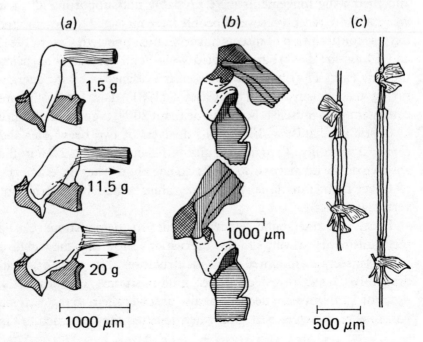

| (a) | (b) | (c) |

involved in the attachment of the wings to the thorax. They consist of layers of resilin interleaved between thin sheets of chitin. In the deformations shown in Fig. 4.5, the resilin is strained much more than the chitin. The elastic tendon of dragonflies (Fig. 4.5(c)) is a cylinder of pure resilin interpolated in the apodemes (the arthropod equivalent of tendons) of some minor wing muscles. The prealar arms and wing hinges of locusts seem to serve as parallel elastic elements (Fig. 4.2(a)) and to be responsible for a substantial fraction of the observed stiffness of the locust wing mounting. The elastic tendons of dragonflies are series elements (Fig. 4.2(b)) and seem unlikely to be important energy stores because they are in series with very minor muscles, not with the principal wing muscles.

The elastic tendon is tiny, less than 1 mm long, but Weis-Fogh (1961b) succeeded in performing tensile tests on it. He found that its Young's modulus was 2 MPa and that it stretched to strains of about 2 before breaking. Jensen & Weis-Fogh (1962) performed dynamic tests on the prealar arm, at a range of frequencies that spanned the natural wing beat frequency, and found that the energy dissipation was small. These are the kinds of properties to be expected of an amorphous, cross-linked polymer in the equilibrium region (section 1.5), and that is what resilin seems to be. They are very like the properties of elastin and abductin.

The low energy dissipation of resilin has often been stressed, but the 3 % energy loss quoted by Jensen & Weis-Fogh (1962) is not the same as the energy dissipation defined in section 1.3. This is because the stress was not taken down to zero in each cycle. The true energy dissipation must have been higher.

4.3 Fibrillar flight muscle

The elastic properties of muscle were introduced in section 1.7. This section shows that the elastic properties of a distinctive type of muscle may have an important energy-saving role in insect flight.

This type is called fibrillar muscle. It forms the main flight muscles of many (but by no means all) groups of insects: it is found in flies, beetles, bugs, bees and wasps, etc. Unlike other kinds of striated muscle, it does not need an action potential to stimulate each contraction. For example, muscles driving the wings of a blowfly at

120 Hz may receive action potentials at a frequency of only 3 Hz (Usherwood, 1975).

Machin & Pringle (1959) demonstrated the mechanical properties of fibrillar flight muscle by experiments on the rhinoceros beetle *Oryctes* and other very large tropical beetles. They used the apparatus shown in Fig. 4.6(*a*). The beetle's body is firmly attached to a force transducer. One of its muscles is exposed and attached by a thin wire to a moving-coil vibrator such as are used to drive loudspeakers. A vane attached to the wire partly blocks a beam of light aimed at a phototransistor (an electrical device sensitive to light). Movements of the vane alter the amount of light falling on the phototransistor and so affect its electrical output. Thus, the force transducer registers forces exerted by the muscle and the phototransistor registers its changes of length. An electrode is used to stimulate the muscle.

The vibrator was used in two different ways. In some experiments it was used to hold the muscle at a series of chosen lengths, at each of which the force was measured. The two broken lines in Fig. 4.6(*b*) were obtained in this way. They are graphs of force against length, the upper one made with the muscle stimulated and the lower one

Fig. 4.6.(*a*) Apparatus used by Machin & Pringle (1959) to investigate the mechanical properties of beetle flight muscle. From R. McN. Alexander (1983). *Animal Mechanics* (2nd edn). Oxford: Blackwell Scientific. (*b*) Graphs showing muscle forces, measured in the experiment illustrated in (*a*), plotted against length. Modified from Machin & Pringle (1959).

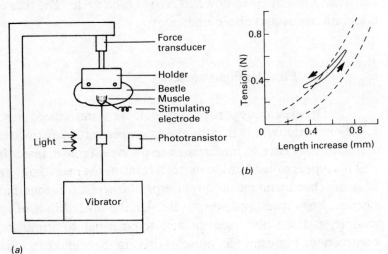

with it unstimulated. These lines show that the muscle has elastic properties. It exerts more force when stimulated, but its stiffness (indicated by the slopes of the lines) is about the same, whether it is stimulated or not.

In the second group of experiments the vibrator was connected to circuits which simulated stiffness and mass. An elastic structure with one end fixed exerts a force proportional to the displacement of the other end, so stiffness was simulated by a circuit that made the vibrator exert a force proportional to the displacement of the vane. Similarly, mass was simulated by a circuit that made the vibrator exert a force proportional to the acceleration of the vane. Machin & Pringle found that when suitable combinations of stiffness and mass were used, stimulation of the muscle set it into oscillatory contraction, making the vane and the coil of the vibrator vibrate up and down. If the simulated mass or the simulated stiffness was changed, the frequency of the contractions changed. The same results would presumably have been obtained if real masses and springs had been used instead of the simulated ones. The simulation was used only for experimental convenience.

This experiment was very like the situation shown in Fig. 4.2(a), in which a muscle with a parallel spring drives a mass. The simulated spring in Machin & Pringle's experiment should be regarded as in parallel with the muscles because spring and muscle experience the same displacements. Equation 4.9 says that the optimum spring stiffness for the parallel-spring arrangement is

$$S_{opt} = m\omega^2 = 4\pi^2 m f^2$$

where m is the mass and $\omega/2\pi = f$ is the frequency. When the spring in Fig. 4.2(a) has this stiffness, it does all the inertial work, leaving the muscle to do only the hydrodynamic work. If there were no hydrodynamic work (and no friction), no work would be required of the muscle: once started, the system would continue vibrating without the need for any action. This can be expressed by saying that a mass m mounted on a spring of stiffness S has a resonant frequency of $\omega_0/2\pi = f_0$

where

$$S = m\omega_0^2 = 4\pi^2 m f_0^2$$
$$f_0^2 = S/4\pi^2 m \tag{4.14}$$
$$f_0 = (1/2\pi)(S/m)^{1/2} \tag{4.15}$$

(The effect of damping on the resonant frequency is ignored here, but is mentioned in section 7.1.)

Equation 4.14 says that for a resonant system with constant mass, f_0^2 is proportional to S. Fig. 4.7(a) shows f^2 plotted against the simulated stiffness in Machin & Pringle's experiments, when the simulated mass was kept constant at 0.03 kg. The graph is a straight line which intersects the stiffness axis at -850 N m^{-1}. Machin & Pringle suggested that the muscle contracted at the resonant frequency of any system in which it was incorporated. Fig. 4.7(a) is consistent with this, if the stiffness of the muscle itself was 850 N m^{-1}. The results of the experiment recorded by the broken lines in Fig. 4.6(b) showed that the stiffness was about that.

Equation 4.14 also says that for a resonant system with constant stiffness, $1/f_0^2$ is proportional to m. Fig. 4.7(b) shows $1/f^2$ plotted against the simulated mass in Machin & Pringle's experiments, for a constant simulated stiffness. Again the results are consistent with the hypothesis that the muscle contracted at the resonant frequency of the system.

The continuous line in Fig. 4.6(b) is a record from one of these experiments. At first sight, it may seem very like the records of tensile tests on tendon shown in Fig. 1.6: like them it is a graph of

Fig. 4.7. Results of the experiment illustrated in Fig. 4.6(a). (a) The relationship between frequency (f) and simulated stiffness, for a simulated mass of 0.03 kg. (b) The relationship between frequency and simulated mass, for a simulated stiffness (additional to the stiffness of the muscle) of 400 N m^{-1}. Modified from Machin & Pringle (1959).

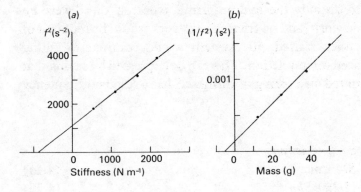

force against displacement that forms a loop. However, there is an important difference. It is an anticlockwise loop, but the loops in Fig. 1.6 are clockwise. Tendon and other passive elastic materials do less work in an elastic recoil than was needed to stretch them. The muscles in Machin & Pringle's experiments did *more* work when shortening than was done on them when they were being stretched. The areas of the loops in Fig. 1.6 represent energy dissipated in the tendon but the area of the loop in Fig. 4.6(*b*) represents work done by the muscle. (This work was done against viscous properties simulated electrically by the circuits connected to the vibrator.) Active fibrillar flight muscle can be thought of as an elastic material with *negative* energy dissipation.

Experiments on active striated muscle from frogs showed that it could stretch elastically only by about 3 %: bigger stretches made cross-bridges detach, losing their strain energy (section 1.7). Fibrillar flight muscles are arranged so that very small length changes suffice to move the wings. That makes it possible for the muscles themselves to serve as energy-saving springs.

Thus, fibrillar flight muscle has its own built-in parallel spring. Insects that have it may not need extrinsic elements, like the resilin springs of locust wing hinges (Fig. 4.5; locust flight muscle is not fibrillar). This idea does not seem to have been worked out quantitatively, for any insect. However, there has been a lot of interest in the elastic properties of the thoraxes of flies (which have fibrillar flight muscles).

4.4 The click mechanism

Boettiger & Furshpan (1952) discovered a phenomenon called the click mechanism, in experiments with flesh flies (*Sarcophaga*). They found that flies anaesthetized with carbon tetrachloride were immobilized with their wings either up or down. Wings that stopped in the up position could be brought down by squeezing the thorax longitudinally. They came down slowly until, when half way down, they clicked suddenly to the fully down position. Similarly, wings in the down position could be brought up by dorso-ventral squeezing of the thorax. They rose slowly until half way, when they clicked suddenly to the fully up position. The wings would not rest in a half-way position, but tended always to move fully up or down. This behaviour is like that of an electric switch which will stay on or off, but not in any intermediate position.

Fig. 4.8(*a*) shows a click mechanism. It is not a literal model of the supposed click mechanism of flies, but shows the basic principle. The compression spring is most compressed when the wing is horizontal but is unloaded as the wing moves up or down, so the wing tends always to click into the up or down position. The graph shows the moments that the spring exerts on the wing: an upward moment when the wing is above the horizontal, and a downward one when it is below the horizontal. In the horizontal position the system is in equilibrium, but it is an unstable equilibrium. The slightest disturbance will make the wings click up or down.

Fig. 4.8(*b*) shows a model with no click mechanism, but with

Fig. 4.8. Diagrams of possible wing mechanisms: (*a*) a click mechanism; (*b*) a mechanism with springs representing wing muscles; and (*c*) a system with both a click mechanism and muscle springs. The graphs show the moments that the springs exert on the wings at different wing angles.

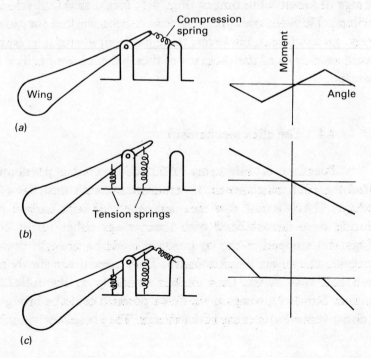

tension springs to represent the elastic properties of the wing muscles. Again, it is not a literal model: the wing muscles of flies are not arranged in the way it implies. However, it will serve our purpose. The two springs together exert a downward moment on the wing when it is above the horizontal and an upward moment when it is below the horizontal. They tend always to bring it back to the horizontal, which is a position of stable equilibrium.

Fig. 4.8(c) shows a model with both sets of springs, representing both the click mechanism and the muscles. The precise behaviour of this model depends on the relative stiffnesses of the springs. One of the possibilities is shown in the graph: the moments from the click mechanism and the muscle springs may be almost equal, cancelling each other out, over much of the range of movement of the wings. In this case the wings would have neutral equilibrium, neither stable nor unstable, when horizontal.

The models shown in Figs 4.8(b) and (c) would both vibrate, when set in motion. The one without the click mechanism (Fig. 4.8(b)) would make simple harmonic motion: a graph of wing angle against time would be a sine wave. The one with the click mechanism (Fig. 4.8(c)) would move differently. It would move at almost constant velocity in the middle part of its range, and decelerate and re-accelerate abruptly at the ends. A graph of wing angle against time would not be a sine wave, but more like a zigzag.

Miyan & Ewing (1985a,b) argued that the click mechanism is an artefact. It works in flies anaesthetized with carbon tetrachloride, which seems to cause excessive contraction of one particular muscle, changing the relative positions of some small parts of the skeleton. It does not seem to work in normal, unanaesthetized flies. Miyan & Ewing devised an explanation of the mechanism of wing movement, with no click action. They based it mainly on detailed study of the anatomy of the wing base, but claimed that it was supported by their films of tethered flies trying to fly. These films show the wings moving at almost constant angular velocity in the middle parts of the downstroke and of the upstroke. Miyan & Ewing argued that, if there were a click mechanism, the wing should accelerate suddenly after passing the horizontal. They seem to have forgotten the elastic properties of the muscles. The wing movements they observed were as predicted for the model shown in Fig. 4.8(c), except that the upstroke was faster than the downstroke.

Their explanation of the wing movements has been challenged by Ennos (1987). Here is a simplified version of his explanation. The wing-bearing segment of the thorax is enclosed mainly by two sheets of stiff cuticle, the dorsal scutum and the ventral sternum (Fig. 4.9). The wings are attached to both but the scutum is narrower, so raising it lowers the wings (Fig. 4.9(*b*)) and lowering it raises them (Fig. 4.9(*a*)). The wings are worked by two sets of muscles, neither of which attaches directly to them. The longitudinal muscles tend to shorten the thorax, when they contract. They make the scutum buckle upwards, and so drive the wings down (Fig. 4.9(*b*)). The dorso-ventral muscles restore the shape of the scutum, raising the wings again. Notice that the scutum is arched transversely, like the roof of a tunnel. When it buckles upwards, arching longitudinally, the transverse arch is flattened at the bend, exaggerating the effect on the wings. The strain energy stored by these distortions of the thorax has not been investigated. We do not know the relative importance of the cuticle and the muscles as springs in the wing mechanism.

Miyan & Ewing's (1985*b*) films of tethered flies showed the upstroke faster than the downstroke, implying that elastic elements were not doing all the inertial work: kinetic energy taken from the

Fig. 4.9. Diagrams showing how the indirect wing muscles of dipteran flies move the wings. The lower diagrams are transverse sections.

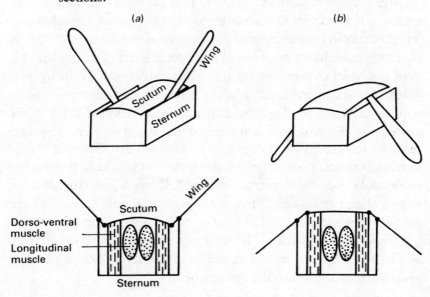

wings at the bottom of the downstroke and stored as elastic strain energy would not be enough to supply the kinetic energy for the upstroke. However, the wing movements may not have been the same as in free flight. Ellington's (1984*a*) films of the fly *Eristalis* hovering show up and down strokes about equally fast.

4.5 Flexible feathers

No energy-saving springs have been identified in the thoraxes of birds. They have nothing like the resilin wing hinges of locusts or the flexible scutum of flies. The tendons of their wing muscles seem too short to be important springs. The muscle fibres lengthen and shorten by about 20 % in each wing beat cycle (Cutts, 1986), which is much more than the 3 % or so that active muscle can stretch elastically (section 1.7). This implies that the strain energy stored in the muscles is probably very small, compared to the work done by the muscles in a wing beat cycle, so muscle elasticity is unlikely to be important.

Pennycuick & Lock (1976) suggested that the wing feathers might be important springs. The primary feathers at the wing tip bend quite markedly during the downstroke, especially in hovering and slow flight (Fig. 4.10(*a*)). They straighten by elastic recoil at the end

Fig. 4.10. Pigeons (*Columba livia*) flying slowly. The front view (*a*) shows how the primary feathers bend during the downstroke. It has been drawn from a photograph in Pennycuick & Lock (1976). The side view (*b*) shows the positions of the wings at the top of the upstroke and the bottom of the downstroke. It is a composite of outlines drawn by Brown (1963) from a cine film.

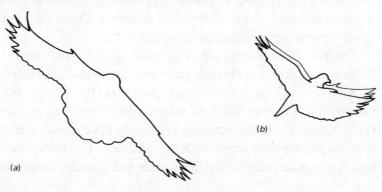

(*a*)   (*b*)

of the stroke, so their tips keep moving while the rest of the wing is decelerating, and the aerodynamic forces on the wing tip are maintained a little longer than if the feathers were rigid. This extra aerodynamic force, late in the stroke, may help to decelerate the wings. Thus, the strain energy stored in the feathers and the kinetic energy of the wing may both be converted to useful aerodynamic work.

The idea applies only to hovering and slow flight because the forward speed of the bird, in fast flight, makes it easy to obtain large aerodynamic forces even at the bottom of the downstroke. Pennycuick & Lock (1976) measured the stiffness of pigeon feathers and made calculations that tended to show that the idea was feasible. However, it seems to me that there is a very serious difficulty.

The aerodynamic forces on a pigeon wing, in the downstroke of slow flight, act almost vertically upwards and bend the feathers upwards (Fig. 4.10(a)). However, the wings beat in a plane that is nearer horizontal than vertical (Fig. 4.10(b)). Only a small component of the aerodynamic force acts in the stroke plane, and can serve to decelerate the wing. It would be particularly small in hovering, when the stroke plane is most nearly horizontal.

4.6   Whales

Whales swim by beating their tails up and down. Every stroke requires hydrodynamic work, and positive and negative inertial work. There is scope for energy saving by springs, just as in insect flight (Bennett *et al.*, 1987).

The springs that seem likely to be important in insect flight act in parallel with the muscles. The springs to be considered in this section act in series with the muscles. The different effects of parallel and series springs were discussed in section 4.1.

The principal swimming muscles of whales, that move the tail fluke up and down, are homologous with the tail muscles of terrestrial mammals. Their muscle fibres are quite short (40–50 mm in a 1.4 m porpoise, *Phocaena*) but their many, slender tendons are very long (200–500 mm). These tendons obviously have elastic compliance, and are arranged in series with the muscles. The vertebral column also has elastic compliance: tension in the muscles compresses the vertebrae and intervertebral discs, shortening the column a little. Paradoxically, this compliance also must be considered to be in series with the muscles: because of it, the muscle fibres must shorten more than would otherwise be necessary, to bring the fluke to a particular position, so its effect is like that of the tendon compliance. Bennett *et al.* (1987) used our dynamic testing machine to investigate the elastic properties of the tendons in tension, and of segments of the vertebral column in compression. We also tested the vertebral column in bending but concluded that it was too flexible to store useful amounts of strain energy by bending. (If the flexural stiffness of the column were important, it would have to be considered, confusingly, as being in parallel with the muscles.)

The inertial forces required for insect flight are needed mainly to accelerate the wings themselves. The added mass of air moving with the wings is trivial. In the case of whale swimming, because water is so much denser than air, the added mass of water is greater than the mass of the fluke.

It is difficult to calculate the forces on the flukes of swimming whales because there is very little quantitative information about their swimming movements. The best data come from analysis of one very short film sequence of a dolphin (*Lagenorhynchus*) swimming at 5.2 m s^{-1} (Lang & Daybell, 1963, run 15.22). They do not include angles of attack, which is unfortunate because changes in angle of attack affect inertial forces. It is not even certain whether the power exerted in that short sequence was precisely the power needed for steady swimming at that speed. Also, there are doubts about the theory used to calculate the hydrodynamic power. For all these reasons, our conclusions admit some doubt, though they seem clear enough to allow a considerable margin for error. The springs seem much less stiff than would be optimal. They could presumably have been made stiffer, by evolving thicker tendons and stiffer intervertebral discs. It remains a puzzle why this has not happened.

Though the possibility has been considered in detail only for whales, other animals that swim by rhythmic movements of hydrofoils may also save energy by elastic storage. They include tunnies and other fishes, turtles and penguins.

Another elastic mechanism has been suggested as a possible energy saver, in the swimming of whales and fishes. It is the subject of the next chapter.

5

Fibre-wound animals

5.1 Helical fibres

Many more-or-less cylindrical animals have helical collagen fibres in their body walls. Some of the fibres run clockwise and some anticlockwise, so that the animal is wrapped in a lattice of fibres (Fig. 5.1(*a*, *b*)). These animals include some nematode and nematomorph worms, which have fibres in their cuticle; planarian and nemertean worms, which have them in the basement membrane of the epidermis; fishes, which have them in their skin; and whales, which have them under their blubber (Clark, 1964; Wainwright, Vosburgh & Hebrank, 1978; Alexander, 1987). Wainwright *et al.* (1978) suggested that the fibres in fish skin might serve as energy-saving springs in swimming. Wainwright, Pabst & Brodie (1985) suggested

Fig. 5.1.(*a*),(*b*) Diagrams of a worm and a fish, showing helically wound fibres. (*c*),(*d*) Diagrammatic cross-sections of a fish and a whale, respectively.

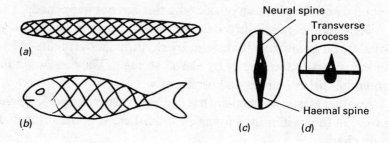

the same, for the fibres under whale blubber. The possibilities are discussed in this chapter.

Harris & Crofton (1957) and Clark & Cowey (1958) made classic analyses of the shape changes of fibre-wound worms. They assumed that the fibres were inextensible and they considered only lengthening and shortening (not bending), so their work is not directly relevant to a discussion of elastic mechanisms in fishes and whales. Nevertheless, a few words about it will help to clarify the rest of the chapter.

Inextensible fibres running lengthwise along a worm will (obviously) prevent it from lengthening. Inextensible fibres running circumferentially around a worm of circular cross-section will prevent it from shortening because it cannot shorten without getting fatter, if its volume is constant. However, a worm with only longitudinal fibres can shorten (if the fibres buckle) and one with only circumferential ones can lengthen (if the fibres buckle, or if the cross-section can flatten to a non-circular shape).

The effect of helical fibres depends on the angle they make with the long axis of the body (α_s, Fig. 5.2(*a*)). If this angle is small, they have essentially the same effect as longitudinal fibres, but if it is near 90°,

Fig. 5.2. Diagrams illustrating the discussion of worm-like animals with helical fibres in the body wall. From R. McN. Alexander (1987). (*J. theor. Biol.*, **124**, 97–110.

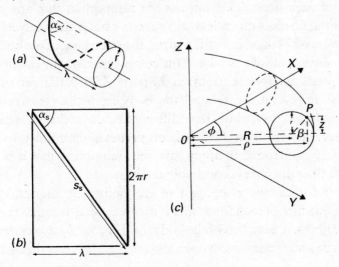

they have essentially the same effect as circumferential ones. Harris & Crofton (1957) showed that worms of circular cross-section with α_s less than 55° can shorten, but not lengthen. Ones with α_s greater than 55° can lengthen, but not shorten. The nematode worm *Ascaris* has helical fibres in its cuticle running at angles α_s of about 76°. Therefore it cannot shorten. If the dorsal longitudinal muscles contract, they bend the worm, stretching the ventral muscles. If the ventral muscles contract, they bend the worm the other way, stretching the dorsal ones.

That theory assumed that the fibres are inextensible. However, collagen fibres are extensible, and can be stretched by up to 8% before breaking (section 1.3). Most of the rest of this chapter is based on a paper that discusses what happens when animals wound in extensible fibres bend (Alexander, 1987).

5.2 A worm-like model

Consider a cylindrical animal of circular cross-section wound by fibres which, when the animal is straight, make helices of angle α_s (Fig. 5.2(a)). Equal numbers of fibres run clockwise and anticlockwise. The animal is very long, and every fibre makes many turns around its body. When the animal bends, fibres may be stretched or allowed to shorten, but the strain is always uniform all along the length of each fibre. This implies the assumption that any matrix around the fibres is much less stiff than the fibres themselves. If the fibres were embedded in a stiff matrix, the stretching of the body wall on the convex side of a bend and the compression on the concave side would produce opposite strains in the parts of fibres that ran through them. A fibre that ran around the body from the concave to the convex side and back would have different strains in different places.

Little is known about the matrix properties in the animals to which the theory is intended to apply, but the assumption that it is much less stiff than the fibres is probably realistic.

Fig. 5.2(a) shows a segment of the body just long enough to include one turn of each fibre: that is to say, its length equals the pitch λ of the fibres. The radius of the body is r. Fig. 5.2(b) shows the body wall of the same segment cut open and laid flat. It shows that the pitch is

$$\lambda = 2\pi r \cot \alpha_s \qquad (5.1)$$

and that the length of fibre needed to make one turn is

$$s_s = 2\pi r \operatorname{cosec} \alpha_s \qquad (5.2)$$

(The subscript in α_s and s_s indicates that these values apply when the body is straight.)

Now let the animal bend into a uniform curve of radius R. This makes it a segment of a torus (a shape like a wedding ring). It can be shown that if its length and volume are unchanged, the radius r is also unchanged. (The length referred to here is measured along the curve of the body.)

Fig. 5.2(c) shows a segment of the bent body in Cartesian (XYZ) space, with the Z-axis along the axis of bending. The position of a point P on the body wall could be described by coordinates (x, y, z), but it is simpler to use the cylindrical coordinates (ϱ, ϕ, z) which are shown on the diagram. The angle β, defined by the diagram, will also be useful. Notice that

$$\varrho = R + r \cos \beta \qquad (5.3)$$

and

$$z = r \sin \beta \qquad (5.4)$$

We can discover the length of a fibre by dividing it into short segments and adding their lengths together. Consider the segment from P (the point ϱ, ϕ, z) to the nearby point $((\varrho + \delta\varrho), (\phi + \delta\phi), (z + \delta z))$. The length δs of that segment is given by

$$\delta s^2 = \delta\varrho^2 + \varrho^2 \delta\phi^2 + \delta z^2 \qquad (5.5)$$

(This is a standard equation for distances measured in cylindrical coordinates.)

To go further, we have to say something about the paths of the fibres in the body wall. They were helices when the body was straight, but what will they be when it bends? If they are taut, they must be geodesics. Just as the shortest line in a plane, joining two points, is a straight line, the shortest line on a curved surface, joining two points, is a geodesic. Geodesics on cylinders are helices but geodesics on tori have no special name. There is a mathematical theorem that says that $\varrho^2 \cdot d\phi/ds$ has the same value, all along a geodesic

$$\varrho^2 \cdot d\phi/ds = C \qquad (5.6)$$

The constant C is different for different geodesics but we can calculate its value for a geodesic that makes one complete turn in a length λ of the torus.

Equation 5.6 can be written

$$\delta\phi = C \cdot \delta s/\varrho^2$$

which can be substituted into equation 5.5, giving

$$\delta s^2 = \delta\varrho^2 + C^2\delta s^2/\varrho^2 + \delta z^2$$
$$\delta s^2(1 - (C^2/\varrho^2)) = \delta\varrho^2 + \delta z^2$$
$$\delta s^2 = \varrho^2(\delta\varrho^2 + \delta z^2)/(\varrho^2 - C^2) \qquad (5.7)$$

Differentiation of equations 5.3 and 5.4 give

$$\delta\varrho = -r \sin \beta \cdot \delta\beta$$

and

$$\delta z = r \cos \beta \cdot \delta\beta$$

Thus,

$$\delta\varrho^2 + \delta z^2 = r^2 (\sin^2 \beta + \cos^2 \beta)\delta\beta^2$$
$$= r^2\delta\beta^2 \qquad (5.8)$$

By substituting this into equation 5.7 we get

$$\delta s^2 = \varrho^2 r^2 \delta\beta^2/(\varrho^2 - C^2)$$
$$\delta s = \varrho r(\varrho^2 - C^2)^{-1/2} \cdot \delta\beta \qquad (5.9)$$

The length s of one turn of the geodesic fibre can be obtained by integrating equation 5.9 around one complete turn (from $\beta = 0$ to $\beta = 360°$). The strain in the fibre, which we want to know, is $(s - s_s)/s_s$. (Its length was s_s when the body was straight.)

Results of such calculations are shown in Fig. 5.3. Notice that all the strains are negative: bending the body does not stretch helical fibres, but allows initially stretched fibres to shorten. The effect is greatest if the angle α_s of the helices is small. Strains are shown for several different radii of curvature R. (Small values of R represent sharp bends.) Cylindrical animals rarely bend so sharply as to make $R/r = 2$, but values of 5 and 10 seem to be common (Fig. 5.3(b)).

This chapter started with the idea that helical fibres may have an energy-saving rôle in swimming, like the rôles of the elastic elements discussed in Chapter 4. The idea was that as a fish beats its tail from side to side, or as a whale beats its fluke up and down, kinetic energy might be converted to elastic strain energy at the end of each stroke, and then recovered in an elastic recoil. For this to work, the strain energy would have to increase as the animal bent and then decrease again as it straightened. The animal would have stable equilibrium in the straight position and would tend to straighten by elastic recoil after being bent.

The model we have just analysed does not behave like that. When it bends, the fibres slacken (Fig. 5.3(*a*)). A straight animal with its fibres taut would be unstable because it could lose strain energy by bending into one or more curves. The fibres could not work in the manner suggested, to save energy in swimming. However, it will be argued in the next section that the model may be inappropriate for fishes and whales, which have anatomical features that have so far been overlooked.

The model may nevertheless help us to understand nematode worms such as *Ascaris* that have helical fibres in their cuticle. Live nematodes are seldom straight, but are usually bent into wavy curves. This may be because the straight position is unstable. Nematodes are admittedly not quite like the model. They have no vertebral column to keep their length constant, so the slackening of the fibres caused by bending would presumably allow the longitudinal muscles (which maintain pressure in the body cavity) to shorten. However the conclusion, that the straight position would be unstable, remains valid. It might be advantageous for a parasitic worm to be bent, if bending helped to anchor it at its preferred site.

Fig. 5.3.(*a*) Strain due to bending of the body, in worm-like animals with helical fibres in the body wall. The strain in the fibres is plotted against initial helix angle for the three radii of curvature shown in (*b*). (*a*) is from R. McN. Alexander (1987). *J. theor. Biol.*, **124**, 97–110.

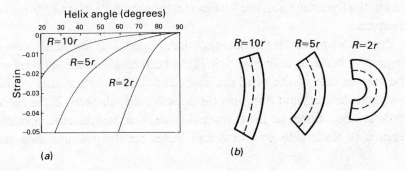

5.3   Fishes and whales

The worm-like model assumed that the fibres ran uninterrupted around the body. This is realistic for nematode worms, but
possibly not for fishes and certainly not for whales. Fishes have
neural spines projecting up from the vertebrae in their tails, and
haemal spines projecting down (Fig. 5.1(c)). Many have their skin
attached to the tips of these processes. Similarly, whales have
transverse processes projecting horizontally from their vertebrae,
with the helical collagen fibres attached to their tips (Fig. 5.1(d)).
Fish, with vertical spines, beat their tails from side to side, but
whales, with horizontal processes, beat their tails up and down.

The attachments to spines or transverse processes divide the fibres
into segments. They separate the segments on the convex side of a
bend from those on the concave side. They may prevent the averaging
out of strains around the body which, in the previous model, gave
each fibre the same strain all along its length. If the spines are very
flexible (as in many fishes), their mechanical effect may be small. If
they are sufficiently stiff, they may make the worm-like model
unrealistic. They seem very stiff in whales.

A second model (Alexander, 1987) has spines that are assumed to
be utterly rigid, with the helical fibres attached to their tips. It is also
assumed that these spines and the membranes between them prevent
tissues and fluids from crossing from one side of the body to the other
(in fishes) or between the dorsal and ventral halves of the body (in
whales). As the animal bends, the half of the body that lies on the
concave side of the bend shortens and gets correspondingly fatter.
The half on the convex side lengthens and gets thinner. If these
changes stretch fibres on the concave side, they allow fibres on the
convex side to slacken, and vice versa. The mathematical analysis is
much like the analysis of the first model (section 5.2), but a little more
complex.

Fig. 5.4 shows length changes calculated for the same three
degrees of bending as in Fig. 5.3. If the helix angle is small, fibres on
the convex side of the bend are stretched (because their side of the
body gets longer) and fibres on the concave side slacken. If the helix
angle is large, fibres on the convex side slacken (because the circumference of their side gets less) and fibres on the concave side are

stretched. Mild bends cause no strain in the fibres on either side of the body if the helix angle is 60°. (Notice that this angle is different from the critical angle of 55° that appeared in the discussion of length changes in section 5.1.) Tendons, and presumably also the collagen fibres in the skin of fishes and under the blubber of whales, break at strains of about 0.08. Fig. 5.4 shows that quite sharp bends should be possible without breaking fibres, for a wide range of helix angles.

If the helix angle is different from 60°, bending increases strain energy on one side of the body and decreases it on the other. If it is greater than 60°, the increase in strain on the concave side (Fig. 5.4(*b*)) is greater than the decrease on the convex side (Fig. 5.4(*a*)). Bending will almost certainly increase strain energy and the animal will tend to straighten by elastic recoil. The proposed energy-saving elastic mechanism is feasible. If the angle is less than 60° the situation is more doubtful. The decrease of strain on the concave side will exceed the increase on the convex side, and strain energy may decrease. However, strain energy may increase if the fibres on the concave side go slack (so that further bending does not decrease their

Fig. 5.4. Strain due to bending of the body in helical fibres that are attached to vertebral processes. Strain is plotted against initial helix angle for the same three radii of curvature as in Fig. 5.3. (*a*) refers to the convex side of the bend and (*b*) to the concave side. From R. McN. Alexander (1987). *J. theor. Biol.*, **124**, 97–110.

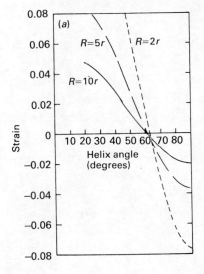
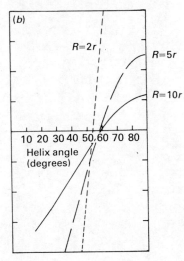

strain energy) or if the fibres are non-Hookean, with their tangent modulus increasing with strain as for tendon collagen (Fig. 1.6). In such cases the energy-saving mechanism would be feasible.

Thus, the mechanism could work, if the helices had any angle other than 60°. The angles observed in fishes are about 60° over most of the body of selachians and 45–50° on the caudal peduncles of selachians and on teleosts (Videler, 1975; Wainwright *et al.*, 1978; Hebrank, 1980). No measurements of the angles in whales seem to have been published.

Real animals are not simple cylinders, as assumed in this chapter. The mathematical arguments that produced Fig. 5.4 assumed that the vertebral processes were utterly rigid, but they may bend appreciably. It was assumed that the axial length of the vertebral column was constant, whereas it may be compressed appreciably by muscle tension. Simplifying assumptions about the shape of the cross-section in bent animals were also made. (Alexander, 1987, discusses them in an appendix.) For all these reasons we should be cautious about applying the mathematical conclusions to real animals. We need experimental confirmation of the predicted strains.

6

Springs as catapults

6.1 Catapults and jumping

A child using a catapult does work stretching the rubber, storing up elastic strain energy. When he fires, the elastic recoil converts much of this strain energy to kinetic energy of the missile. The child does work on the catapult and gets much of it back. If that were all there were to it, there would be no point in using catapults: it would be just as effective to throw the stone. The important point about catapults is that they throw stones faster than the unaided arm could do. Work is done slowly on the catapult as the rubber is stretched and returned rapidly in the elastic recoil. The work returned is inevitably a little less than the work previously done, but the rate of working (the power) is greatly increased. Catapults are power amplifiers.

Catapults throw stones but jumping animals throw themselves. We will see that some of them use the catapult principle, but first we have to discuss the limits to jumping ability. The length and height of a jump depend on the speed and angle to take-off. For example, an animal of mass M that takes off vertically with velocity v has kinetic energy $\frac{1}{2}Mv^2$ at take-off. It will rise to height h, acquiring potential energy Mgh, where g is the gravitational acceleration. At that height the initial kinetic energy has been converted to potential energy

$$Mgh = \tfrac{1}{2}Mv^2$$
$$h = v^2/2g \tag{6.1}$$

Similarly, it can be shown that an animal taking off with velocity v from level ground can make a jump of length v^2/g if it takes off at 45° to the horizontal. (At any other angle it travels less far.) These calculations have ignored energy losses due to air resistance but are not seriously inaccurate except for large jumps by very small animals such as fleas (Bennet-Clark & Alder, 1979).

Small jumpers have another problem, as well as air resistance. Most of them jump by extending their legs, like the flea shown in Fig. 6.1. They have to accelerate from rest to their take-off velocity in a distance approximately equal to the length of their legs. Call this distance l. If the animal accelerates uniformly from rest to velocity v over this distance, it covers the distance at a mean velocity $v/2$ and the time required is

$$t = 2l/v \tag{6.2}$$

To jump to height h it must take off with velocity $(2gh)^{1/2}$ (equation 6.1) and

$$t = 2l/(2gh)^{1/2} = l(2/gh)^{1/2} \tag{6.3}$$

Smaller animals have shorter legs than large ones and so have shorter acceleration distances l. For a given height, their acceleration times are shorter. For example, the flea *Spilopsyllus* accelerates to its

Fig. 6.1. Superimposed outlines of a flea (*Spilopsyllus*) about to jump and (broken outline) taking off drawn from high-speed cine films by Bennet-Clark & Lucey (1967). The flea is about 1.5 mm long. From R. McN. Alexander (1979). *The Invertebrates.* Cambridge: Cambridge University Press.

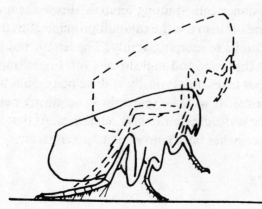

take-off velocity of 1 m s^{-1} over a distance of only 0.4 mm (Bennet-Clark & Lucey, 1967). Equation 6.2 says that its acceleration time must be a mere 0.8 ms, and high-speed cinematography confirms it. Compare it with a much larger jumping animal, the bushbaby *Galago senegalensis*. In a particularly impressive jump, a bushbaby took off at 6.7 m s^{-1} and reached a height of 2.3 m (Hall-Craggs, 1965). Its acceleration distance was 16 cm, so the acceleration time, by equation 6.2, was 48 ms. If the two animals had taken off at equal velocities, the disparity between their acceleration times would have been even greater. Small animals have to accelerate to take-off velocity in very short times.

It seems unlikely that any muscle could make a single complete contraction in the flea's acceleration time of 0.8 ms. A mosquito (*Aëdes aegypti*) beating its wings at 600 Hz (Weis-Fogh, 1973) makes each wing stroke in 0.8 ms, but its wing muscles are not making an isolated contraction: they are in the peculiar state of oscillation which is the special property of insect fibrillar muscle. The flea does not, in fact, contract its jumping muscles in 0.8 ms. It contracts them much more slowly, using them to strain blocks of resilin in a catapult mechanism (Bennet-Clark & Lucy, 1967). The movement that occurs in 0.8 ms is the elastic recoil of the resilin.

6.2 Locusts' jumping

Catapult mechanisms have been found in all the insects that jump well that have been investigated (Bennet-Clark, 1975). I have chosen to describe the mechanism found in locusts (*Schistocerca gregaria*) because it was studied particularly thoroughly by Bennet-Clark (1975). Locusts jump much like fleas (Fig. 6.1) by extending their long hind legs. Films showed that they took off at speeds up to 3.2 m s^{-1}. This would be enough for a 1.0 m long jump if the take-off angle were optimal and if air resistance were negligible: the longest jumps observed were about 0.95 m. The kinetic energy of a 1.7 g locust taking off at 3.2 m s^{-1} is $\frac{1}{2} \times 0.0017 \times (3.2)^2 = 0.0087$ J, so the work each hind leg had to do was 4.4 mJ.

The films showed that the acceleration time was 25–30 ms. This is much longer than for fleas (because the locusts have longer legs), and is long enough for a mechanism without a catapult to be feasible. A catapult is nevertheless used.

Locust legs have two long segments, called the femur and tibia, which meet at a joint called the knee (Fig. 6.2). Each segment is a tube of fairly stiff cuticle. The tibia is slender but the femur is stout and contains the muscles principally responsible for the jump. These are the large extensor muscle which extends the knee and the small flexor which bends it. The extensor muscle powers the jump and the flexor has a rôle in the catapult machanism, as will be explained. Each muscle is attached to the tibia by an apodeme made of similar material to the outer cuticle. On each side of the knee, the femur has a thickening of its cuticle called the semilunar process. The tibia is hinged to the distal ends of these processes.

Bennet-Clark (1975) measured the forces the two muscles could exert when they were stimulated electrically. He removed a leg from a locust and pinned it to a block of wood. The apodeme of one of the muscles was tied by thread to a force transducer and the muscle was stimulated by electric shocks applied through the pins. He found that the extensor muscle in male locusts could exert a maximum of 13 N, and the flexor only 0.7 N. The force in the extensor muscle

Fig. 6.2. The knee of the hind leg of a locust (*a*) intact and (*b*) dissected to show the muscles. From H. C. Bennet-Clark (1975). *J. exp. Biol.*, **63**, 53–83.

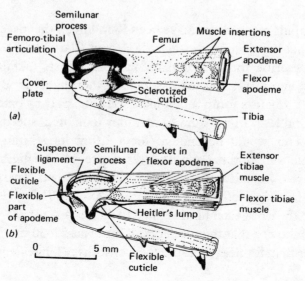

rises and falls quite slowly at the beginning and end of stimulation: at 30 °C it needs about 0.35 s to rise to its maximum and 0.12 s to fall to zero. The forces in the flexor muscle rise and fall much faster.

When an intact leg with its knee bent is stimulated electrically, the knee remains bent. Both muscles are stimulated but the small flexor is able to prevent the large extensor from extending the knee, because it has a much larger moment arm. The dot labelled 'femoro-tibial articulation' in Fig. 6.2(*a*) marks the axis of the joint. Comparison with Fig. 6.2(*b*) shows that the line of the extensor apodeme passes very close indeed to the axis of the joint, much closer than that of the flexor apodeme. In addition, the structure called Heitler's lump forms a catch that helps to keep the knee bent (Heitler, 1974).

The knee remains bent while the stimulus continues but extends rapidly when the stimulus ends. This is because the force declines more rapidly in the flexor than in the extensor, releasing the catch and allowing the extensor to extend the knee.

When the leg is stimulated, the semilunar processes can be seen to distort, like a fist being clenched more tightly. They are distorted by tension in the muscles, and strain energy is stored in them. The tension must also store strain energy in the apodemes. Bennet-Clark (1975) argued that the semilunar processes and the extensor apodeme were the main strain energy stores in a catapult mechanism. They are represented as springs in Fig. 6.3. Tension in the extensor muscle compresses the spring that represents the process and stretches the one that represents the apodeme, storing strain energy in both. When the flexor relaxes, both springs recoil and extend the knee. The diagram illustrates another feature of the real knee: the moment arm

Fig. 6.3. Diagrams of the locust knee (*a*) bent with the muscle taut and (*b*) extended. The extensor apodeme and semilunar processes are represented as springs.

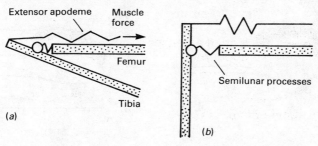

of the extensor increases as the knee extends. Although the force in the apodeme decreases during the elastic recoil, the increasing moment arm makes the force on the ground rise to a peak.

Fig. 6.4 shows how Bennet-Clark measured the strain energy stored in the semilunar processes. The femur was pinned to a wooden block and the extensor muscle was stimulated. The force on the tibia, needed to prevent the knee extending, was measured and used to calculate the force in the extensor. In the experiment illustrated, a force of 13 N caused 0.4 mm distortion. If the spring were Hookean, the strain energy would be 2.6 mJ (by equation 1.4). However, Bennet-Clark found that the system was not Hookean and that the strain energy (measured as the area under a graph of force against displacement) was a little higher.

He calculated the strain energy in the apodeme from the results of an experiment in which he bent apodemes by pushing them against a piece of hairspring from a watch. He measured the deflection of the apodeme and calculated the force from the deflection of the hairspring. By using the theory of bending beams, he calculated that Young's modulus for the apodeme was 19 GPa. He then used this result and the dimensions of the apodeme to calculate that the force exerted by the extensor muscle would store 3 mJ strain energy in the apodeme. Later, Ker (1980) used a miniature testing machine to make tensile tests on the apodeme and confirmed that Bennet-Clark's modulus was about right.

Fig. 6.4. Drawings traced from photographs of an experiment to measure strain energy storage in the semilunar processes. The extensor muscle is stimulated in (*b*) but not in (*a*). Notice the movement of the articulation relative to the fixed datum. From H. C. Bennet-Clark (1975). *J. exp. Biol.*, **63**, 53–83.

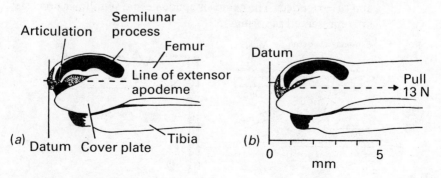

These experiments showed that the semilunar processes of each hind leg could store about 3 mJ strain energy and the apodeme a further 3 mJ, giving a total for the leg of 6 mJ. We calculated at the beginning of this section that the jump requires 4.4 mJ from each leg. The catapult stores more than enough energy to power the jump.

An intriguing feature of the locust's catapult is that it seems only just strong enough to withstand the forces of jumping. If the semilunar processes are scratched with the point of a scalpel, they break when the locust next tries to jump (Bennet-Clark, 1975). The extensor muscle can break its own apodeme, and sometimes does so if the leg is stimulated while fully flexed. However, it seems to be rare for a locust to break its leg by jumping. This may be because the forces of jumping are very predictable. I discussed this point when I formulated a theory of the safety factors of skeletons (Alexander, 1981).

6.3 Heavy feet

To jump, a locust extends its hind legs, moving them relative to the rest of its body. Therefore, it has internal as well as external kinetic energy, when it leaves the ground. Its muscle must do enough work to supply both. The internal kinetic energy is small for the locust because the mass of its legs (and especially of their distal parts) is small.

Fig. 6.5(a) shows a simple model designed to illustrate the principle. It shows an animal of total mass M, consisting of a body of mass $(1 - a)M$ connected by a spring to a foot of mass aM. (a can be any fraction between 0 and 1.) Initially, the spring is compressed, with

Fig. 6.5. A model of a jumping animal with a heavy foot.

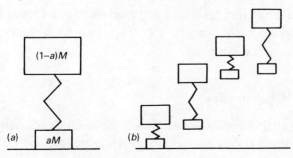

strain energy E stored in it. It is released and accelerates the body to velocity V. If all the strain energy is converted to kinetic energy

$$E = \tfrac{1}{2}(1 - a)MV^2 \qquad (6.4)$$

At the instant when the strain energy reaches zero, the body (mass $(1 - a)M$) has velocity V and the foot (mass aM) is still stationary. The velocity of the centre of mass is the weighted mean of these two velocities: it is $(1 - a)V$. The external kinetic energy is half the mass times the square of this velocity

$$\text{external kinetic energy} = \tfrac{1}{2}M(1 - a)^2\,V^2$$
$$= (1 - a)E \qquad (6.5)$$

(using equation 6.4). The internal kinetic energy at this instant is the difference between this and the total kinetic energy given by equation 6.4

$$\text{internal kinetic energy} = aE \qquad (6.6)$$

Only the external kinetic energy is converted to potential energy. This is because the decelerating effect of gravity acts uniformly on the whole animal. It reduces the upward velocity of the centre of mass, so that external kinetic energy is lost as potential energy is gained, but it does not affect movements of parts of the body relative to the centre of mass. If the external kinetic energy $(1 - a)E$ is converted to potential energy Mgh, raising the centre of mass to height h

$$(1 - a)E = Mgh$$
$$h = (1 - a)E/Mg \qquad (6.7)$$

and the smaller a is (the lighter the foot) the higher the jump.

As the model leaves the ground, the spring lifts the foot. It is stretched by the force required and the system starts oscillating, with the spring alternatively stretching and recoiling as the model flies through the air (Fig. 6.5(b)). This oscillation involves energy aE (equation 6.6) shuttling back and forth between two forms: internal kinetic energy and strain energy. If the aim is to jump as high as possible, this energy is wasted.

6.4   Click beetles

That analysis will throw light on the jumping of click beetles (*Elateridae*), though they jump without using their legs.

Evans (1972) studied *Athous haemorrhoidalis*, which is 10–12 mm long. He found these beetles on humid summer days clinging to the tops of grass blades. When he tried to catch them, they often dropped to the ground. If they landed upside-down and could not right themselves quickly, they jumped, sometimes to a height of 0.3 m.

Athous and other click beetles jump from a lying position by suddenly bending their backs at the joint between two segments of the thorax. Fig. 6.6 has been traced from a film, taken with a special camera at the very high speed of 3100 frames per second. Initially,

Fig. 6.6. Outlines traced from successive frames of a film, taken at 3100 frames per second, of *Athous* taking off for a jump. The numbers are the serial numbers of the frames indicated. The scale lines are marked at intervals of 1/10 inch (2.5 mm). From M. E. G. Evans (1972). *J. Zool.*, **167**, 319–36.

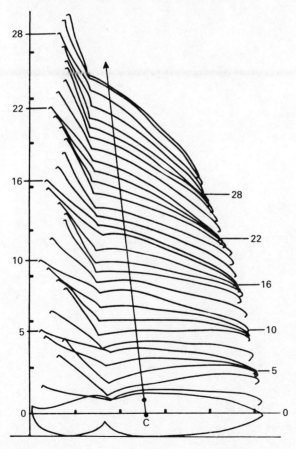

the beetle was laying along the line 00 with its centre of mass at C. It bent its back, raising its centre of mass and throwing itself into the air. The film shows it accelerating to take-off in just two frames, or 0.6 ms. This is even shorter than the acceleration time of the flea discussed in section 6.1. Evans (1972) showed that this very rapid movement was brought about by a catapult mechanism. Contraction of a large muscle strains parts of the cuticle of the thorax. A catch mechanism allows the energy to be released very rapidly.

Fig. 6.6 shows that the beetle oscillates, bending and extending its back as it flies through the air. This happens for the same reason as the oscillation of the model in Fig. 6.5, and like it wastes energy. Evans (1973) calculated that one-third of the energy used for jumping was lost in this way.

7

Suspension springs and shock absorbers

7.1 Suspension systems

Fig. 7.1(a) is a diagram of the suspension system of a car. Each wheel is attached to the body by a spring and a shock absorber. The elastic properties of the spring allow the wheels to move up and down relative to the body as the car drives over bumps and potholes, so that the passengers get a reasonably smooth ride. The shock absorbers make vibrations die quickly away so that the car does not continue bouncing up and down, after going over a bump (Wong, 1978).

Fig. 7.1(b) shows a simple type of shock absorber, a dashpot. It is a cylinder filled with viscous oil, with a leaky piston. To move the piston, oil must be forced through holes in it or through the narrow gap around it. The force needed to do this, because of the viscosity of the oil, is proportional to the velocity $\dot{x}$ of the piston relative to the cylinder

$$F = K\dot{x} \tag{7.1}$$

where K is the viscous damping constant. Work done by this force is degraded to heat and cannot be recovered in the way that work can be recovered from a spring in an elastic recoil.

In Fig. 7.1(a) the sprung mass is the mass of the car body and the unsprung masses are the masses of the wheels. The spring and dashpot shown under each unsprung mass represent the elastic and viscous properties of the tyres.

Fig. 7.1.(*a*) A diagram of the suspension system of a car. From J.
Y. Wong (1978). *Theory of Ground Vehicles.* New York: Wiley. (*b*)
A diagram of a dashpot. (*c*),(*d*) The model discussed in the text.

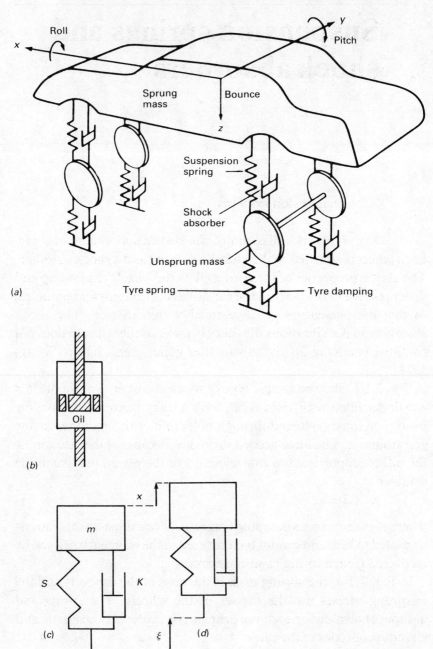

This chapter is about mechanisms in animals that are in some way analogous to the suspension systems of cars. To get some basic understanding of such mechanisms we will use the simple system shown in Fig. 7.1(c). This resembles the suspension system of one wheel of the car, with the mass of the wheel and the properties of the tyre ignored. It consists of a mass m with a spring of stiffness S and a dashpot of damping constant K. In Fig. 7.1(c) the system is in equilibrium but in Fig. 7.1(d) it has been disturbed. The foot has been pushed up by a height ξ and the mass has risen by a height x. The upward velocities of the foot and mass are $\dot{\xi}$ and $\dot{x}$.

The spring has lengthened by $(x - \xi)$ so an upward force $S(x - \xi)$ must act on its top. The dashpot is lengthening at a rate $(\dot{x} - \dot{\xi})$ so an upward force $K(\dot{x} - \dot{\xi})$ must be acting on it. The mass has acceleration $\ddot{x}$ so an upward force $m\ddot{x}$ must be acting on it. However, no external force is acting on the mass. Therefore

$$S(x - \xi) + K(\dot{x} - \dot{\xi}) + m\ddot{x} = 0 \tag{7.2}$$

To get a simple situation that is reasonably easy to analyse, let the foot move up and down sinusoidally

$$\xi = a \sin \omega t \tag{7.3}$$
$$\dot{\xi} = a\omega \cos \omega t \tag{7.4}$$

Assume that, in response, the mass moves up and down with the same frequency but not necessarily the same phase

$$x = b \sin (\omega t + \delta) \tag{7.5}$$
$$\dot{x} = b\omega \cos (\omega t + \delta) \tag{7.6}$$
$$\ddot{x} = -b\omega^2 \sin (\omega t + \delta) \tag{7.7}$$

By substituting equations 7.3 to 7.7 into equation 7.2 we get

$$S[b \sin (\omega t + \delta) - a \sin \omega t]$$
$$+ K\omega[b \cos (\omega t + \delta) - a \cos \omega t] - m\omega^2 b \sin (\omega t + \delta) = 0 \tag{7.8}$$

We want to know how large the movements of the mass are in comparison with the movements of the foot. In other words, we want to know the transmissibility ratio b/a. Notice that the second term in equation 7.8 is multiplied by ω and the third by ω^2. At very low frequencies these terms will be small compared to the first one, so

$$S[(b \sin (\omega t + \delta) - a \sin \omega t] \approx 0 \tag{7.9}$$

which can only be true if $b \approx a$ and $\delta \approx 0$. This means that the mass moves up and down in phase with the foot and with the same amplitude, as if it were attached rigidly to the foot. At very high frequencies the last term in equation 7.8 would be much larger than the others and the equation could not balance, unless b were very small. This means that the mass would stay almost stationary, unaffected by the movements of the foot. The first term in equation 7.9 has the factor S and the final term has the factor $m\omega^2$. Therefore the changeover from the low-frequency situation ($b \approx a$) to the high-frequency situation (b declining towards zero) occurs when S has the same order of magnitude as $m\omega^2$: that is, when ω has values near $(S/m)^{1/2}$ (i.e. at frequencies near $(1/2\pi)$ $(S/m)^{1/2}$). In this range of frequencies, very large transmissibilities occur if K is small: the body may bounce up and down far more than the foot. The frequency at which b/a is highest is called the resonant frequency. It is approximately $(1/2\pi)(S/m)^{1/2}$ if K is small, and less if K is large.

Fig. 7.2 shows how the transmissibility ratio depends on the frequency and on the damping constant. (The frequency ratio is the frequency divided by $(1/2\pi)(S/m)^{1/2}$.) The main effect of damping is to reduce the height of the peak at the resonant frequency.

Fig. 7.2. Graphs of transmissibility ratio b/a against frequency ratio $\omega/(S/m)^{1/2}$ for the system shown in Fig. 7.1(c),(d). Each line refers to a different damping constant K and is labelled with the appropriate value of $K/[2(mS)^{1/2}]$. From J. Y. Wong (1978). *Theory of Ground Vehicles*. New York: Wiley.

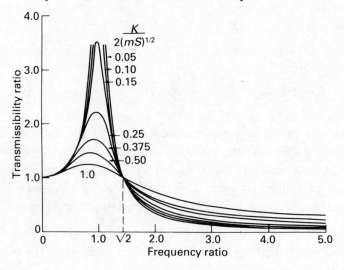

7.2 Loads on the head

African women habitually carry loads on their heads, whether the load be a container of fruit or water, a bundle of firewood or almost anything else. In Kenya, women of the Luo tribe carry the load on top of the head but Kikuyu women carry it hanging behind the back from a strap that runs over the forehead (Maloiy *et al.*, 1986). These loads may be as much as 70 % of the woman's own mass. The habit of carrying things on the head is not restricted to Africans or to women: for example, the (male) porters in Covent Garden market, London, used to carry baskets of fruit and vegetables on their heads.

Maloiy *et al.* (1986) wondered whether less energy was needed to carry loads on the head, than in other ways. They measured the rates of oxygen consumption of Luo and Kikuyu women, carrying loads in their traditional ways. The women walked on a moving belt so that they were stationary relative to the laboratory. The air they breathed out was collected through a mask and analysed to discover how much oxygen had been removed from it. Fig. 7.3 shows some of the results and compares them with similar data for soldiers carrying loads in back packs. When they walked without loads, the women and the soldiers used oxygen at approximately equal rates per unit body mass, at all speeds. When they carried 34 kg loads their rates of oxygen consumption were higher, but the increase was much smaller for the women than for the soldiers, at most speeds. Despite their different load-carrying techniques, Luo and Kikuyu women use oxygen at about equal rates.

These data seem to show that loads are carried more economically on the head than on the back by people accustomed (in each case) to carry loads that way. Why should that be? Maloiy and his colleagues suggested that loads on the head might travel almost horizontally, as if on wheels, instead of rising and falling with the body. Thus, the need to do work to increase their potential energy during each step would be avoided. Any potential energy fluctuations of the load might be partially balanced by kinetic energy fluctuations, using the pendulum principle that is also used when walking without loads (Alexander, 1986) but it might still be an advantage to keep the load on a level course.

This could be done by a spring that would be compressed at the stage of the step when the body was high (Fig. 7.4(*a*, *c*)) and extended when it was low (Fig. 7.4(*b*)) (Alexander, 1986). The diagrams show an external spring on top of the head, but the elastic properties of the back and neck may have the required effect. The back and neck would have to behave as passive springs: it would be no use using muscle activity to keep the load level, because the muscles would then have to do positive work between stages (*a*) and (*b*) (when the back would have to extend while exerting force) and negative work between (*b*) and (*c*). The elastic properties of tendons and ligaments and of intervertebral discs might enable the back to function as a suitable spring.

The frequency would have to be high enough to give a low transmissibility ratio. Fig. 7.2 suggests that it should be at least twice

Fig. 7.3. Rates of oxygen consumption per unit body mass for people walking at different speeds, with and without loads. $\triangle$, $\bigcirc$, Luo and Kikuyu women, respectively, without loads; $\blacktriangle$, $\bullet$, the same, carrying 34 kg loads on their heads; lower broken line, soldiers without loads; upper broken line, soldiers with 34 kg loads in back packs. Bars show 95 % confidence intervals. From G. M. O. Maloiy *et al.* (1986). *Nature*, **319**, 668–9.

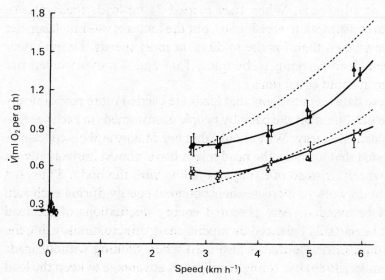

the resonant frequency

$$\omega/(S/m)^{1/2} \geqslant 2$$
$$\omega^2 \geqslant 4S/m$$
$$S \leqslant m\omega^2/4 \tag{7.10}$$

where S is the stiffness of the spring, m is the mass of the load and ω is 2π times the frequency of the vertical movements of walking. If the spring were Hookean, setting the load on the head would compress it, reducing the carrier's height by

$$\Delta h = mg/S$$
$$\geqslant mg/(m\omega^2/4)$$
$$\geqslant 4g/\omega^2 \tag{7.11}$$

Walking people take about two steps per second, so the body rises and falls at a frequency of about 2 Hz. Thus, ω is about 4π s^{-1}. The gravitational acceleration g is 9.8 m s^{-1}. Thus, Δh would have to be at least $4 \times 9.8/16\pi^2 = 0.25$ m. The back would have to be compliant enough for the head to be pushed down this far by the weight of the load. It seems unlikely that it could be, and photographs of people carrying head loads suggest that it is not. This does not necessarily mean that the head is not mounted on a spring that gives low transmissibility: the spring might be non-Hookean, or its unloaded length might be adjusted by muscle tension. However, it makes it seem rather unlikely.

Fig. 7.4. Diagrams of a person carrying a load on the head.

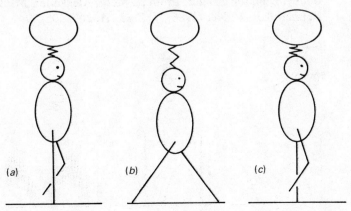

7.3 Paw pads

Fig. 7.5(*a*) shows again the simple model of running that was introduced in section 3.4. All the mass is supposed to be concentrated in the block representing the trunk; the mass of the spring (representing the leg) is negligible. When the model bounces, the force on the ground rises smoothly to a peak and falls again, as shown in the graph at the bottom of the figure.

That model is unrealistic because real feet have mass. Fig. 7.5(*b*) is a revised model with a small mass (representing the foot) on the lower end of the spring. When it hits the ground, the foot mass is decelerated very rapidly to rest and a large transient force acts, as shown in the graph. This might cause damage.

Most mammals have fatty pads in their paws, that cushion the impact with the ground. Account of this has been taken in Fig. 7.5(*c*) by adding a second spring under the mass of the foot, to represent the pad. When this model bounces, there are oscillations superimposed on the main rise and fall of the ground force, because the impact sets the foot mass vibrating up and down on the pad. Records of the forces exerted by people and animals running over force plates do not show continuing vibrations like this, so the model needs further refinement.

Fig. 7.5. Diagrams of models referred to in the discussion of paw pads and (below) graphs showing the force each would exert when bouncing on the ground. From R. McN. Alexander, M. B. Bennett & R. F. Ker (1986). *J. Zool., A*, **209**, 405–19.

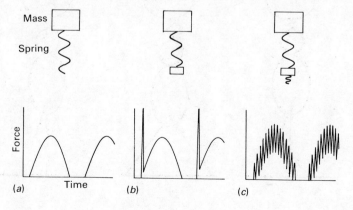

Fig. 7.6(a) shows a slightly more complex model that was analysed by Alexander, Bennett & Ker (1986). The paw pad now incorporates a dashpot as well as a spring. This model resembles a one-wheeled version of the car shown in Fig. 7.1 (only the shock absorber is missing). When this model hits the ground, the initial vibration dies away, because of the damping (Fig. 7.6(b)). This graph is very like the force plate record of a barefoot human runner, shown in Fig. 7.7(a).

We will enquire what properties a paw pad should have, to prevent the peak force following impact (f_{max}, Fig. 7.6(b)) being too large, in comparison with the peak reached in the middle of the step (F_{max}). We will estimate f_{max}/F_{max} making three assumptions that seem likely to be realistic.

(i) The mass m of the foot is small, compared to the mass M of the body.

(ii) The stiffness s of the paw pad is large, compared to the stiffness S of the leg. This seems likely because paw pads are quite thin (15–20 mm in large dogs) and no force, however large, can compress them by more than a fraction of their thickness.

(iii) The peak force F_{max} is large, compared to body weight. It is about three times body weight for people running at moder-

Fig. 7.6.(a) The model of an animal with a paw pad that is analysed in detail. (b) A schematic graph showing the force it would exert after landing on the ground. From R. McN. Alexander, M. B. Bennett & R. F. Ker (1986). *J. Zool.*, A, **209**, 405–19.

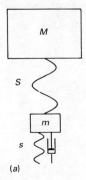
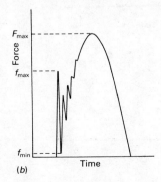

ate speeds (Fig. 7.7) and up to six times body weight for kangaroos hopping fast (Alexander & Vernon, 1975).

Let the model fall vertically onto the ground. At the instant of impact the springs are unstrained, the body is falling with velocity V and the foot is falling with velocity v. (V and v are not necessarily equal.)

A mass m mounted on a spring of stiffness s has a resonant frequency $(1/2\pi)(s/m)^{1/2}$ (section 7.1), so its period of vibration is $2\pi(m/s)^{1/2}$. If the foot were detached from the body and leg, it would come to rest before accelerating upwards again in about one-quarter of a period, $(\pi/2)(m/s)^{1/2}$. The mean deceleration is the initial velocity divided by this time, $(2v/\pi)(s/m)^{1/2}$. The mean force is the mass multiplied by the deceleration, $(2v/\pi)(sm)^{1/2}$. A similar argument for the body says that the mean force needed to decelerate it is $(2V/\pi)(SM)^{1/2}$. The peak forces should be in the same ratio as the means, so

$$f_{\max}/F_{\max} \approx (v/V)(sm/SM)^{1/2} \tag{7.12}$$

Fig. 7.7. Records of the vertical component of force exerted by people running across a force plate. (*a*) A man running barefoot. (*b*) The same man, wearing jogging shoes. (*c*) A man running barefoot, using a midfoot striking technique. From J. A. Dickinson, S. D. Cook & T. M. Leinhardt (1985). *J. Biomechan.*, 18, 415–22.

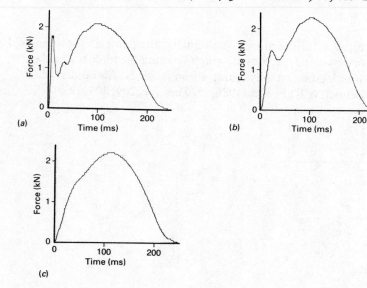

(This argument depends on assumption (iii). If the peak force on foot or body were not large compared to the weight, the deceleration time would be more than a quarter cycle.) This shows that a light foot (small m) with a soft pad (small s) brought down slowly (small v) will exert only a small impact force f_{max}. Comparison of Fig. 7.7(a) and (b) shows that the extra padding (lower s) provided by a shoe reduced f_{max} for a runner.

There seems to be another requirement for a satisfactory foot pad, in addition to making f_{max} reasonably small. Fig. 7.6(b) has been drawn with the first minimum of force (f_{min}) greater than zero. However, it is conceivable that it (and some of the subsequent minima) should reach zero. In other words, the foot might bounce up and down, losing and regaining contact with the ground, before finally settling. We call this phenomenon 'chattering' and suppose it would be disadvantageous, because the foot would be apt to shift position each time it left the ground.

How can chattering be prevented? If the foot were separated from the body and leg, it would bounce off the ground after half a cycle of vibration, at time $\pi(m/s)^{1/2}$. Its velocity then would be less than v because of the damping effect of the dashpot: call it ev. (e is called the coefficient of restitution and may have any value between 1 for no damping and 0 for very strong damping.) The momentum of the foot would be mev and it would leave the ground (chattering would occur) unless this momentum had by then been removed by the leg pressing down on it. The necessary impulse (mean force × time) would be mev.

In this early part of the step, the leg spring S is being compressed at a rate V. After time t it has been compressed by a distance Vt, and the force on it is SVt. The mean force up to this time is $\frac{1}{2}SVt$ and the impulse is $\frac{1}{2}SVt^2$. The time in question is $\pi(m/s)^{1/2}$ (see above), so this impulse is $\frac{1}{2}SV\pi^2m/s$. The condition for no chattering is that this should be at least mev

$$\frac{1}{2}SV\pi^2m/s \geqslant mev$$
$$s/S \leqslant \pi^2V/2ev$$
$$\leqslant 5V/ev \qquad (7.13)$$

Alexander *et al.* (1986) checked this crude analysis by numerical simulation and found that its predictions were reasonably accurate, provided there was not too much damping (e was not too low).

We now have two requirements for a satisfactory paw pad. The ratio of peak forces (f_{max}/F_{max}) should not be too large, and chatter should not occur. We will find out which is easier to satisfy and then concentrate on the other. If chatter is just prevented, v/V equals $5S/es$ (inequality 7.13). By putting this value into equation 7.12 we get

$$f_{max}/F_{max} \approx (5S/es)(sm/SM)^{1/2}$$
$$\approx (5/e)(Sm/sM)^{1/2} \tag{7.14}$$

Assumptions (i) and (ii) (above) say that the square root term must be small. Therefore f_{max}/F_{max} will not be large unless e is very small. A foot that did not chatter would be unlikely to suffer excessive impact forces unless its pad were very heavily damped.

Thus, the difficult problem seems to be to avoid chatter, and we will concentrate on that. Inequality 7.13 says that chatter could be prevented by heavy damping of the pad (low e) but it will be shown that e is not small. Chatter could be avoided by making v/V small, by decelerating the foot so that it was almost stationary before hitting the ground. However, films of a kangaroo hopping showed that the feet generally came down faster than the body (v was larger than V). Finally, chatter could be avoided by keeping s/S sufficiently small, by having a soft pad. However I argued, in presenting assumption (ii), that s/S must be large. The dilemma can be avoided by having a non-Hookean pad, whose stiffness increases as the force on it increases. Such a pad would have low effective stiffness early in the step (when the danger of chatter arises) but high stiffness later. Numerical simulations show that this could work.

Alexander *et al.* (1986) tested paw pads to discover their properties. We dissected them from carcases of dogs, a kangaroo, a camel and several other mammals. We squeezed them between steel plates in a dynamic testing machine, making the actuator move up and down at various frequencies. Results for a dog pad are shown in Fig. 7.8. The results for the other species were similar.

Notice that the loops in Fig. 7.8(*b*) are curved, indicating low stiffness at low forces and higher stiffness at higher forces. This is the kind of non-Hookean behaviour needed to prevent chatter. It is not a surprising property to find. Any material compressed to large strains can be expected to show increasing stiffness because it cannot be squeezed to zero thickness. The peak forces in Fig. 7.8 are approximately the peak forces that would have acted on the foot in fast trotting or slow galloping (Jayes & Alexander, 1978).

Notice also that the loops in Fig. 7.8(*b*) are not very wide. The energy dissipations are about 25 %, corresponding to coefficients of restitution of about 0.87.

Fig. 7.8.(*a*) Print of a hind paw of a 26 kg dog. (*b*) Records from compressive tests on the metatarsal pad, at two frequencies. These records were made by Dr R. F. Ker, using the method of Alexander, Bennett & Ker (1986).

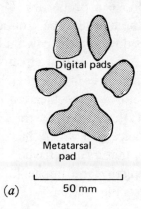

Digital pads

Metatarsal
pad

(*a*) 50 mm

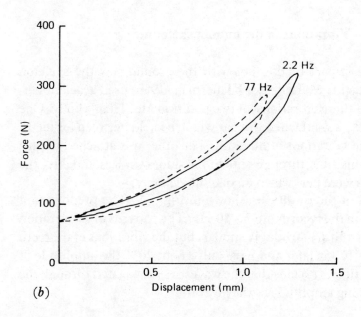

(*b*)

The results are similar for all frequencies, which is convenient because the pattern of force on the foot in running has two frequency components. In Fig. 7.7(a), for example, the main rise and fall of the curve is half a cycle of frequency 2 Hz, but the initial oscillation has a frequency of about 50 Hz. The first is the frequency of mass M vibrating on spring S (Fig. 7.6(a)) and the second is that of mass m vibrating on spring s. In force plate records of dogs galloping, the higher frequency is of the order of 100 Hz. The tests shown in Fig. 7.8 span much of the relevant range of frequencies.

Alexander *et al.* (1986) found it difficult to explain why chatter did not occur, even with the observed non-Hookean properties. One possibility is that muscles may shorten as the foot hits the ground, making the force with which the leg presses down on the paw build up faster than the simple model suggests. Another is that effective pad stiffness may be reduced by setting down the toes first, so that the digital pads (Fig. 7.8(a)) make contact before the metacarpal or metatarsal pads. Films show that dogs and kangaroos do this. It reduces impact forces by increasing the distance over which the paw is decelerated. Similarly, Fig. 7.7(c) shows that a runner who sets down the ball of his foot first avoids producing the large impact force that results from the heel being set down first (Fig. 7.7(a)).

7.4   Vibrations in the human skeleton

The impact of the foot with the ground sets the skeleton vibrating. Light, McLellan & Klenerman (1980) used accelerometers to record these vibrations, at two points in McLellan's body. One accelerometer was attached to a bar which he held between his teeth, to record the vibrations of his skull. The other was attached to pins driven into his tibia through slits in his skin. (Understandably, the experiments were performed on one subject only.)

Examples of the results are shown in Fig. 7.9. The predominant frequencies in the records are 20–50 Hz. The patterns of acceleration of tibia and teeth are strikingly similar, but the vibrations of the teeth occur about 10 ms later and have only about 20 % the amplitude of those of the tibia. It seems that the waves are propagated through the skeleton, losing amplitude as they travel.

All the records were made while walking on a hard laboratory floor. The largest accelerations occurred when McLellan walked barefoot or in shoes with hard leather heels. Both the amplitude and the frequency were reduced when he wore crêpe-soled shoes, or shoes with a plastic shock-absorbing insert in the heel. This seems consistent with the observations of ground forces in running with and without shoes (Fig. 7.7).

The phenomenon illustrated by Fig. 7.9 seems to be the propagation of a wave of elastic strain (that is, a sound wave) through the body. The distance between the attachment points of the accelerometers seems to have been about 1.2 m, so the delay of 10 ms indicates a speed of 120 m s^{-1}.

Sounds travel through solids with velocities of $(E/\varrho)^{1/2}$, where E is the Young's modulus and ϱ the density of the solid. The path from tibia to teeth must pass through materials ranging from compact bone to the much softer tissue of intervertebral discs, materials which differ greatly in modulus and density. For a rough estimate of the expected velocity, assume that the whole path is through cancellous bone like that of the vertebral centra: this material has properties intermediate between those of compact bone and intervertebral discs. Human cancellous bone commonly has Young's modulus of the order of 5×10^7 Pa (Currey, 1984) and a density of about 1500 kg m^{-3}. Thus, the velocity of sound through it should be of the order of $(5 \times 10^7/1500)^{1/2} = 180$ m s^{-1}. This is the same order of magnitude as the observed velocity.

Fig. 7.9. Simultaneous records from accelerometers attached to the tibia and held between the teeth, of accelerations following heelstrike when walking barefoot and with three kinds of shoe. The axial component of acceleration for the tibia, and the vertical component for the teeth, are shown. From L. H. Light, G. E. McLellan & L. Klenerman (1980). *J. Biomechan.*, **13**, 377–80.

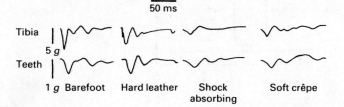

7.5   Breathing while running

The abdominal viscera of mammals (the stomach, intestines, liver, etc.) are not rigidly attached to the body wall. When the animal breathes in, the lungs enlarge, the diaphragm flattens and the viscera move posteriorly. When it breathes out, the lungs get smaller and the viscera move forward again. These movements are driven by the muscles of the diaphragm and (in heavy breathing) the abdominal wall.

When the animal runs, its body accelerates and decelerates in every stride. Consequently, the inertia of the viscera tends to move them forward and back in the trunk. This suggests the model of the trunk shown in Fig. 7.10. Mass m represents the viscera. It is connected to the body wall by a spring of stiffness S and a dashpot of damping constant K. The spring represents the elastic properties of the diaphragm and abdominal wall. The dashpot represents the damping effects of the tissues and of the work needed to pump air in and out of the lungs. Distances ξ and x give the positions of the trunk and the viscera, relative to a fixed point outside the body. If $(x - \xi)$ increases, the viscera move forward in the trunk and the animal breathes out. If $(x - \xi)$ decreases, the animal breathes in.

This model is essentially the same as the model shown in Fig. 7.1(c, d). It has a resonant frequency of about $(1/2\pi)(S/m)^{1/2}$, if K is

Fig. 7.10. A model to show how breathing movements could be driven by the accelerations and decelerations of running.

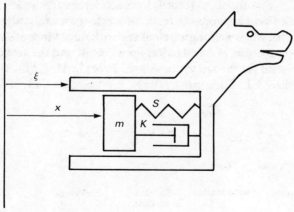

not too large. Fluctuations of ξ tend to cause fluctuations of x, with transmissibility ratios that depend on the frequency.

Each of the four legs of a quadruped tends to decelerate and re-accelerate the body in every stride (section 3.1). The pattern of deceleration and acceleration of the body may therefore be complicated, but it repeats regularly at the stride frequency. If this frequency were much less than the resonant frequency, the transmissibility ratio would be 1 (Fig. 7.2). The viscera would move as if rigidly attached to the body wall, unless driven by breathing muscles. There would be no tendency for running movements to effect breathing. If, however, the stride frequency were much higher than the resonant frequency, the transmissibility ratio would be very low. The viscera would move forward at almost constant velocity, unaffected by the accelerations and decelerations of the trunk. The trunk would alternately move forward relative to the viscera, enlarging the lungs, and drop back again, compressing them. The running movements could power breathing.

Running movements could also power breathing, or at least assist it, if the stride frequency were close to the resonant frequency, giving a transmissibility ratio greater than 1 (Fig. 7.2). In this case the velocity of the viscera ($\dot{x}$) would fluctuate more than the velocity of the trunk ($\dot{\xi}$). There would be an advantage in breathing at the resonant frequency, like the advantage insects get by beating their wings at the resonant frequency of the thorax (section 4.2): no inertial work would be required.

Bramble & Carrier (1983) made experiments to investigate the relationship between running and breathing movements. They attached small stereophonic cassette recorders to dogs, horses and people. One channel of the recorder was connected to a miniature microphone mounted on a face mask. The other was connected to an accelerometer fastened to a foot. The microphone recorded the sounds of breathing and was more sensitive to breathing out than to breathing in. The accelerometer registered the foot's impacts with the ground. Fig. 7.10 shows part of the record of an experiment with a horse. Notice that the cycles of leg movements and of breathing had the same frequency. Horses and dogs kept the two frequencies identical, both in trotting and in galloping. People usually took two strides per breath.

Horses and other quadrupeds use about the same stride frequency at all galloping speeds: they speed up by taking longer strides, not by increasing the frequency (Heglund, Taylor & McMahon, 1974). The stride frequency is little different in fast trotting, but is lower in slow trotting. Horses sometimes breathed erratically when trotting, in

Fig. 7.11. Leg movements and breathing of a cantering horse. (*a*) Some stages of the cycle of leg movements, which are indicated by the same numbers below (*b*), which is the output of an accelerometer attached to the right forefoot. (*c*) The output of a microphone mounted on a face mask showing the sounds of expiration E and inspiration I. (*d*) An approximate representation of the total vertical component of force on the two forefeet. The lines in (*e*) show when each foot was on the ground. RF, LF, right and left forefeet; RH, LH, right and left hind feet. From D. M. Bramble & D. R. Carrier (1983). *Science*, **219**, 251–6.

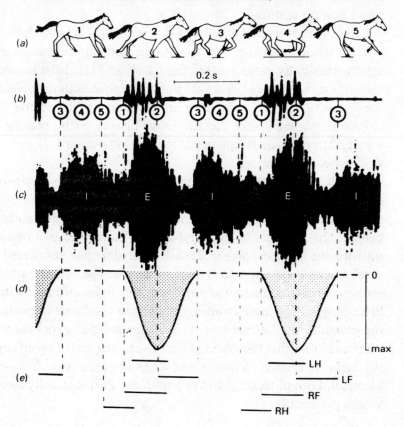

Bramble & Carrier's (1983) experiments, possibly because the stride frequency had dropped too far below the resonant frequency.

These data seem to favour the idea that mammals tend to breathe at the resonant frequency of the respiratory system, and to match their stride frequency to the resonant frequency. However, other data suggest a different resonant frequency for the respiratory system.

Mammals and birds pant to cool themselves. If panting needed a lot of energy to drive it, it would increase metabolic heat production and defeat its own object. Crawford (1962) suggested that they minimize the energy cost by panting at the resonant frequency of the respiratory system. He found that dogs panted at very constant frequencies. The 12 kg dogs that he studied panted at a mean frequency of 5.3 Hz. He anaesthetized the same dogs and put them in a mechanical respirator, which he ran at various frequencies. He found that their thoraxes resonated at a mean frequency of (again) 5.3 Hz. That seems good evidence that dogs pant at a resonant frequency, but it also raises a difficulty. The same dogs would have galloped with stride frequencies of only 3.2 Hz (calculated from an equation given by Heglund *et al.*, 1974). Can the respiratory system have two resonant frequencies, one for panting and the other for breathing while galloping?

The answer is yes, if it has two modes of vibration. The breathing movements we have been discussing involve backward and forward movement of the viscera. Panting seems to involve vibration in and out of the rib cage. (Crawford, 1962, recorded panting movements, by means of an accelerometer fixed to the ribs.) There is no necessity for the two modes of vibration to have the same resonant frequency. Indeed, Crawford's (1962) work suggested that there might be a second resonance. He took X-ray pictures of the dogs immediately after they had been killed at the end of the experiment. He took these pictures with the dog tilted head up and head down and observed how far the diaphragm was moved by the weight of the viscera. Hence he calculated that the resonant frequency for forward and backward vibration of the viscera should be about 4 Hz, which is well below the resonant frequency measured in the respirator.

8

Springs and control

The human hand is much too small to contain all the muscles needed to work it. The muscles it does contain are called the intrinsic muscles. They are numerous but small, and are concerned largely with movements such as spreading the fingers and bringing the thumb into opposition with the fingers. The other, extrinsic, muscles are in the forearm. They include the largest muscles of the hand, the principal flexors and extensors of the fingers and thumb. They are attached to the bones of the hand by long tendons.

These tendons must stretch when their muscles exert forces. Does this make it difficult to control hand movements? Is the elastic compliance of tendons, so useful in other contexts (Chapter 3), an embarrassment here?

Imagine that you are using a finger to hold a small lever in a constant position. You will depend largely on the sense organs called muscle spindles which lie among the muscle fibres and sense any changes of length. Changes from the desired length elicit correcting reflexes. How well can this work if you have to resist forces that are trying to move the lever? Even if your reflexes keep the length of the muscle fibres precisely constant, the fluctuating forces will stretch the tendon and cause movement of the finger.

As an example, consider the long flexor muscle of the thumb, which was studied by Rack & Ross (1984). It bends both joints of the thumb. Its belly, in the forearm, consists of muscle fibres which are 40–50 mm long when the thumb is extended. These insert in pennate fashion on a tendon which runs 100–130 mm along the belly of the muscle and extends a further 100–130 mm beyond the belly, into the hand. Rack & Ross (1984) made tensile tests on the part of the tendon that lies distal to the muscle belly and found that its stiffness at high stresses was about 90 N mm^{-1}. The muscle is capable of exerting forces up to at least 130 N (calculated from Fig. 3D of Brown, Rack & Ross, 1982) which would stretch the tendon by 1.4 mm, enough to allow the distal joint of the thumb to rotate 11°. Even if the spindle reflexes could keep the muscle fibre lengths precisely constant, fluctuating loads would cause appreciable movement at the joint.

The reflexes do not work perfectly, however. They do not work instantaneously, so movements are corrected only after a delay. Fig. 8.1 is a model of the control system. The muscle fibres are represented by a spindle, a spring, a dashpot and a force generator in parallel. The spring in series with them represents the tendon. Length changes are sensed by the spindle, which sends a signal to the force generator, which (after a delay) exerts an appropriate force. We will consider how a system like this is affected by the elastic compliance of the tendon.

Fig. 8.2(*a*) shows a muscle resisting a force which varies sinusoidally about a mean value $\overline{F}$. The value of the force at time t is

$$F = \overline{F} + F_0 \sin \omega t \qquad (8.1)$$

This fluctuating force makes the length of the muscle fluctuate with

Fig. 8.1. A diagram of a muscle and its tendon, illustrating the discussion of stretch reflexes. From P. M. H. Rack, H. F. Ross, A. F. Thilman & D. K. W. Walters (1983). *J. Physiol.*, **344**, 503–24.

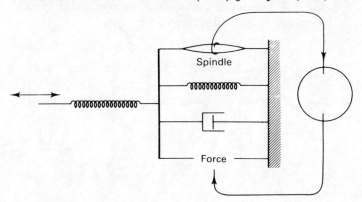

the same frequency, despite the control exerted by the reflex. At time t the deviation of the length from its mean value is

$$x = x_0 \sin (\omega t - \delta) \qquad (8.2)$$

The phase shift δ is due to the delay in the reflex arc and to the viscous properties of the muscle (represented by the dashpot in Fig. 8.1).

Fig. 8.2(b) shows the same muscle with a tendon in series with it. The force on the end of the tendon fluctuates according to equation 8.1, making the length of (muscle plus tendon) fluctuate according to the equation

$$X = X_0 \sin (\omega t - \varepsilon) \qquad (8.3)$$

We will calculate the amplitude X_0 and the phase shift ε, and so discover the effect of the tendon.

Length changes of the tendon are represented by $(X - x)$. If the tendon is perfectly elastic and has stiffness S, the force F will make its length vary according to the equation

$$X - x = (F_0 \sin \omega t)/S$$
$$X = x + (F_0/S) \sin \omega t \qquad (8.4)$$

Values for X and x are given by equations 8.3 and 8.2. By putting them in equation 8.4, we get

$$X_0 \sin (\omega t - \varepsilon) = x_0 \sin (\omega t - \delta) + (F_0/S) \sin \omega t \quad (8.5)$$

Now we have to use the standard equation for the sine of the difference of two angles on both sides of the equation

$$X_0 (\sin \omega t \cos \varepsilon - \cos \omega t \sin \varepsilon)$$
$$= x_0 (\sin \omega t \cos \delta - \cos \omega t \sin \delta) + (F_0/S) \sin \omega t \quad (8.6)$$

Fig. 8.2. Diagrams of muscles (a) without and (b) with tendons, stretched by fluctuating forces F.

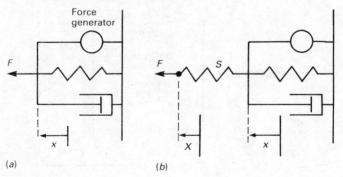

This equation must balance for all values of t. When $t = 0$, $\sin \omega t = 0$ and $\cos \omega t = 1$, so the equation is

$$X_0 \sin \varepsilon = x_0 \sin \delta \tag{8.7}$$

When $\omega t = \pi/2$, $\sin \omega t = 1$ and $\cos \omega t = 0$

$$X_0 \cos \varepsilon = x_0 \cos \delta + (F_0/S) \tag{8.8}$$

By dividing equation 8.7 by equation 8.8

$$\tan \varepsilon = x_0 \sin \delta/[x_0 \cos \delta + (F_0/S)] \tag{8.9}$$

Also, by squaring the two equations and adding them together

$$X_0^2 \sin^2 \varepsilon + X_0^2 \cos^2 \varepsilon$$
$$= x_0^2 \sin^2 \delta + x_0^2 \cos^2 \delta + (F_0/S)^2 + (2F_0x_0/S) \cos \delta$$
$$X_0^2 = x_0^2 + (F_0/S)^2 + (2F_0x_0/S) \cos \delta \tag{8.10}$$

We will apply these equations to the results of experiments with the long flexor muscle of the thumb. The apparatus is shown in Fig. 8.3. The subject's forearm and hand are held immobile in a plaster cast which leaves the distal joint of the thumb free. A lever which is tightly bound to the thumb is connected by a system of levers to a crank on a motor-driven flywheel. Thus, rotation of the flywheel alternately extends the thumb and allows it to bend. The force on the thumb is measured by the force transducer and its position is monitored by a photoelectric device involving the cathode ray tube.

Fig. 8.3. Apparatus for an investigation of the control of thumb movements. From T. I. H. Brown, P. M. H. Rack & H. F. Ross (1982). *J. Physiol.*, **332**, 69–85.

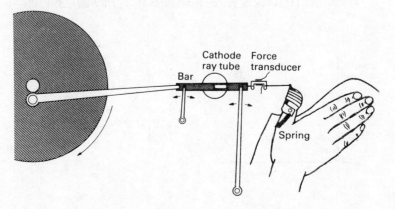

The spring prevents backlash. The extensor muscles of the thumb have been anaesthetized by an injection so that only the flexor is active.

Rack & Ross (1984) used this apparatus to investigate the reflexes of the long flexor muscle. The subject tensed the muscle, maintaining a constant mean force on the transducer while the crank made the thumb vibrate through a small angle. The force fluctuated in a way which depended partly on the passive properties of the tissues and partly on the reflex activity of the muscle.

The records of the experiments gave force F (equation 8.1) and displacement X (equation 8.3) as functions of time. Fig. 8.4(a) shows how the results were displayed on polar plots. The radius F_0/X_0 represents the ratio of force amplitude to displacement amplitude and the angle ε represents the phase lag. Thus, the horizontal axis gives $(F_0/X_0) \cos \varepsilon$, which is called the elastic stiffness, and the vertical axis gives $(F_0/X_0) \sin \varepsilon$, which is called the viscous stiffness.

Results from a typical experiment are shown by the triangles in Fig. 8.4(b). Both the stiffness F_0/X_0 and the phase lag ε vary with frequency. The stiffness S of the tendon had been measured, on

Fig. 8.4.(a) Diagram showing how the frequency responses of control mechanisms are displayed in polar plots. (b) Polar plot showing reflex responses of the interphalangeal joint of the thumb (triangles) and the calculated responses of the muscle fibres (crosses). Stiffnesses are expressed as angular stiffnesses, that is as the torque needed to move the joint through unit angle. From P. M. H. Rack & H. F. Ross (1984). *J. Physiol.*, **351**, 99–100.

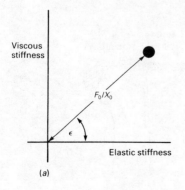
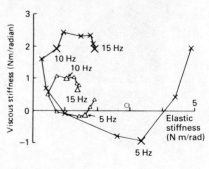

specimens taken from cadavers, and is indicated by a circle near the horizontal axis. Thus, the stiffness F_0/x_0 and phase lag of the muscle fibres could be discovered, by using equations 8.9 and 8.10.

The crosses in Fig. 8.4(*b*) show the calculated behaviour of the muscle fibres. At low frequencies the muscle fibres by themselves would have been much stiffer (would have resisted movement better) than the muscle–tendon combination. At higher frequencies they resisted movement less well, in some cases no better than the muscle–tendon combination.

The spiral shape of the curves in Fig. 8.4(*b*) can be explained by a modified version of the model shown in Fig. 8.1. It is necessary to suppose that there is a delay in the reflex, that the reflex becomes ineffective at high frequencies and that the damping constant of the dashpot falls as frequency increases (Rack, Ross, Thilmann & Walters, 1983). The delay in the reflex, and its ability to operate at high frequencies, explain why the stiffness of muscle fibres is high only at low frequencies.

Fig. 8.4(*b*) shows that the compliance of the tendon greatly reduces the effective stiffness of the muscle fibres at low frequencies. It makes the muscle much less able to hold the thumb in position against a fluctuating force, than if its fibres had been attached directly to the skeleton. This may seem to be a disadvantage.

However, not all the thumb's tasks require precise control of position. Some (such as holding delicate objects) require control of the force it exerts. A compliant tendon makes it harder to control position but it may make it easier to control force, by reducing the force change accompanying a small movement.

The tendon of the long flexor of the thumb is non-Hookean, like other tendons (Fig. 1.6). Its stiffness is low when the force in it is small, but increases with increasing force. Sometimes we may want to control the positions of our thumbs very precisely while exerting little force. We can increase the effective stiffness of the tendons to make this easier, by contracting the flexor and extensor muscles simultaneously. Thus, both may exert large forces on their tendons though the thumb may be pressing only lightly on the object being held.

9

Springs and size

9.1 Elastic similarity

Why have buffaloes got relatively shorter legs than gazelles, and why have albatrosses relatively longer wings than sparrows? Many biologists have asked questions like these, about how the proportions of animals change to suit different body sizes (Schmidt-Nelsen, 1984). One of the concepts they have found most stimulating is that of elastic similarity, as introduced by Rashevsky (1960) and developed by McMahon (1973, 1975).

The idea is that organisms of different sizes may be constructed so as to bend in similar fashion under their own weight, and so as to be equally safe from collapse by buckling. A few examples will help to explain this.

Fig. 9.1(a) shows the neck of an imaginary animal. In the absence of gravity it would be straight, but it bends under the weight of the head. It has length l and diameter d. It is pulled down a distance s by the weight W of the head. Think of it (unrealistically) as a homogeneous cylinder of material of Young's modulus E. A standard equation for the bending of beams says

$$s = Wl^3/3EI \tag{9.1}$$

where I is the second moment of area of the cross-section, which is $\pi d^4/64$ for a cylinder of diameter d (Alexander, 1983, equation 4.3 and Fig. 4.8). Thus,

$$s = 64Wl^3/3\pi E d^4$$

and

$$s/l = 64Wl^2/3\pi Ed^4 \tag{9.2}$$

Elastically similar animals would bend in similar fashion: that is, they would have equal values of s/l. Equation 9.2 tells us that for this to be true, they must have equal values of Wl^2/d^4

$$Wl^2/d^4 = \text{constant} \tag{9.3}$$

(This assumes that they are made of the same material and so have equal Young's moduli E.)

Fig. 9.1(b) shows an animal with legs which will be considered (unrealistically) as homogeneous cylindrical columns made of material of Young's modulus E. Each leg has length l and diameter d and has to support a weight W. A standard equation tells us that a column of these dimensions would buckle and collapse under a load $\pi^2EI/4l^2$, where I is the second moment of area of the cross section, $\pi d^4/64$ (Alexander, 1983, p. 147). If the supported weight is a fraction a of the load that would cause collapse

$$\begin{aligned} W &= a\pi^2EI/4l^2 \\ &= a\pi^3Ed^4/256l^2 \\ a &= 256l^2W/\pi^3Ed^4 \end{aligned} \tag{9.4}$$

Fig. 9.1. Diagrams referred to in the discussion of elastic similarity.

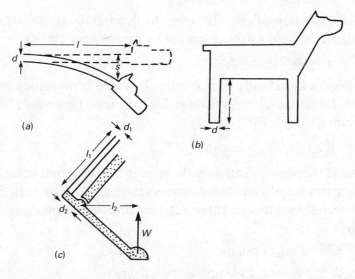

Elastically similar animals would be equally safe from collapse: that is, they would have equal values of a. Equation 9.4 tells us that that would be true only if they had equal values of l^2W/d^4.

$$l^2W/d^4 = \text{constant} \tag{9.5}$$

which is identical with equation 9.3. Our discussions of horizontal necks and vertical legs have led to identical relationships between length, diameter and weight.

Now consider the slightly more realistic leg shown in Fig. 9.1(c). This leg has to support a weight W. The joint is held at a particular angle, and prevented from collapsing under the load by the force transmitted by a tendon of length l_1, diameter d_1. These dimensions are obviously a length and a diameter. It is less obvious that distance l_2 should be regarded as a length and distance d_2 as a diameter. By taking moments about the joint, the force in the tendon is

$$\text{force} = Wl_2/d_2 \tag{9.6}$$

The stress in the tendon is this force divided by the cross-sectional area, $\pi d_1^2/4$

$$\text{stress} = 4Wl_2/\pi d_1^2 d_2. \tag{9.7}$$

The strain is the stress divided by Young's modulus E

$$\text{strain} = 4Wl_2/\pi E d_1^2 d_2 \tag{9.8}$$

and the extension is the strain multiplied by the length of the tendon

$$\text{extension} = 4Wl_1 l_2/\pi E d_1^2 d_2 \tag{9.9}$$

This extension allows the joint to bend through an angle α, calculated by dividing the extension by the moment arm d_2

$$\alpha = 4Wl_1 l_2/\pi E d_1^2 d_2^2 \tag{9.10}$$

The joints of elastically similar animals would bend through equal angles. Equation 9.10 says that for this to be true, they would have to have equal values of $Wl_1 l_2/s_1^2 d_2^2$,

$$Wl_1 l_2/d_1^2 d_2^2 = \text{constant} \tag{9.11}$$

This is the same as equations 9.3 and 9.5, except that in this case we have two lengths and two diameters instead of one of each. Thus, it has been shown that for three different situations, elastic similarity requires

$$Wl^2/d^4 = \text{constant} \tag{9.12}$$

where W is a weight, l a length and d a diameter.

It has been assumed implicitly that all parts of animals scale according to the same rules. Thus, the weight of each part is the same fraction of body weight, in all the animals being compared. Also, ratios like l_1/l_2 and d_1/d_2 (Fig. 9.1(c)) are the same for all. This assumption is often false: for example, skeletons tend to be a larger fraction of body mass in larger mammals (see Hokkanen, 1986). However, we will see where it leads us. If corresponding parts of the body have the same density, in different animals, their weights are proportional to length multiplied by cross-sectional area and so to ld^2. If our assumption is true, the weight of the body or of any part of it is proportional to ld^2, even when the length l and diameter d are those of other parts of the body,

$$W/ld^2 = \text{constant.} \tag{9.13}$$

For equations 9.12 and 9.13 both to be true, the animals must have

$$l \propto W^{0.25} \text{ and } d \propto W^{0.375} \tag{9.14}$$

(You can check this by substituting $l = k_1 W^{0.25}$ and $d = k_2 W^{0.375}$ in equations 9.12 and 9.13) Thus, elastically similar animals have corresponding lengths proportional to (body mass)$^{0.25}$ and corresponding diameters to (body mass)$^{0.375}$. The hypothesis, that terrestrial animals of different sizes tend to be elastically similar, has the beauty of being simple and general. It seemed extremely attractive when it was first published. However, it also has unattractive features. One of them is that it makes distinctions between lengths and diameters which sometimes seem artificial. It was not immediately obvious, in Fig. 9.1(c), that l_2 should be regarded as a length and d_2 as a diameter. An even more awkward anomaly arose in a discussion of running, where it was necessary to treat stride length as a diameter (McMahon, 1975).

Another unattractive feature of the hypothesis is that it is not apparent why terrestrial animals should tend to be elastically similar. They do not, in general, seem to have any undesirable tendency to sag under their own weight. Their legs seem to be in no danger of buckling under the weight of the body. If body proportions had evolved in response to the tendency to sag or buckle under gravity, we might expect to find that the dangers of sagging and buckling were not very remote. It might be argued that animals have evolved to avoid undue sagging or buckling, not under the weight of the standing animal but under the much larger forces that act in running.

If the peak forces involved in running were the same multiple of body weight, for animals of different sizes, the argument leading to the proportionalities in 9.14 could stand. However, the multiples are generally smaller for larger animals. The peak force on the hind foot of a 2.4 kg hare, running at top speed, is about 4.3 times body weight. That on the hind foot of a 2500 kg elephant, also running at top speed, is only about 1.0 times body weight (Alexander, 1985).

The elastic similarity hypothesis has generally been applied only to terrestrial animals, because aquatic ones are supported largely by buoyancy. It predicts that terrestrial mammals will have lengths proportional to (body mass)$^{0.25}$ and diameters to (body mass)$^{0.375}$ (proportionalities 9.14). Alexander (1982) surveyed the available data concerning external dimensions of mammals, and bone dimensions of mammals and birds. Most of the data showed lengths and diameters both about proportional to (body mass)$^{0.35}$. Thus, lengths scale very differently from the prediction of elastic similarity. However, one group of mammals, the ungulates, had body lengths, limb lengths and bone lengths about proportional to (body mass)$^{0.25}$. An example of the data is shown in Fig. 9.2. Notice that the points for lengths of the humerus of Bovidae (antelopes, etc.: ungulate mammals) lie on a line of smaller gradient than the line for mammals in general.

Fig. 9.2. Graph on logarithmic coordinates showing the lengths and diameters of the humerus of mammals plotted against body mass. ○, insectivores; ●, primates; ◇, rodents; ◆, carnivores; □, Bovidae; ■, other Artiodactyla; △, others. From R. McN. Alexander, A. S. Jayes, G. M. O. Maloiy & E. M. Wathuta (1979). *J. Zool.*, **189**, 305–14.

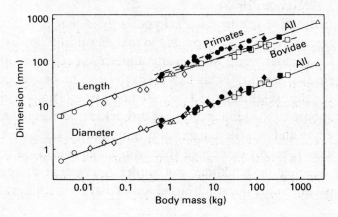

Elastic similarity requires animals of different sizes to have different shapes. Lengths increase more slowly than diameters, with increasing body mass. Therefore, small animals should be more slender in build, and large animals stockier, as in the comparison at the beginning of this chapter between gazelle and buffalo. Notice that these mammals both belong to the Bovidae, the group in which the clearest tendency to elastic similarity has been found.

If animals or other structures of different sizes have the same shape, they are said to be geometrically similar. They must have lengths and diameters both proportional to (body mass)$^{0.33}$. The data reviewed by Alexander (1982) show that the bodies and skeletons of mammals of different sizes (except ungulates) are much closer to geometric than to elastic similarity.

It seemed very odd that the ungulates should be different from the rest of the mammals, until Economos (1982) pointed out that most of the very large mammals, in the data I reviewed, were ungulates. He suggested that the difference was not between ungulates and the rest, but between large mammals and small ones. He showed that (head plus body) length was proportional to (body mass)$^{0.31}$ for mammals in general. However, for mammals smaller than 20 kg it was proportional to (body mass)$^{0.34}$ and for mammals larger than 20 kg it was proportional to (body mass)$^{0.27}$. The smaller mammals scaled approximately as required for geometric similarity and the larger ones as for elastic similarity. It remains a puzzle why this should be.

Thus, the elastic similarity hypothesis is based on unconvincing theory but seems quite good at predicting the scaling of body and bone dimensions in ungulates, and perhaps in large mammals in general.

9.2 Dynamic similarity

The dynamic similarity hypothesis is another that has been adopted in recent discussions of animal sizes. It is not a direct rival to the elastic similarity hypothesis, because it is concerned with movement rather than with structure. Indeed, it is largely compatible with the elastic similarity hypothesis.

The concept of dynamic similarity may be regarded as an extension of that of geometric similarity. Two shapes are geometrically similar if one could be made identical to the other by a uniform change in the scale of length. For example, an animal that was geometrically similar to another might be twice as long, twice as wide and twice as high. Its bones would be twice as long and have twice the diameter. Its hairs would be twice as long and spaced twice as far apart. Two movements are dynamically similar if one could be made identical to the other by uniform changes in the scales of length, time and force. For example, if the animals just described moved in dynamically similar fashion, all the distances involved in their movements would be in the same proportion as the lengths of their bodies: the larger one would take strides twice as long, or jump twice as high. Also, the time taken for every stage of movement would be multiplied by another constant factor. For example, the larger animal might take 1.4 times as long for each stride, and 1.4 times as long to prepare for a jump. Finally, all forces would be multiplied by a third constant factor. The larger animal might be eight times as heavy and exert eight times the force with each muscle.

It should be obvious why I have chosen the factor eight for forces. If an animal is twice as long, twice as wide and twice as high as another of the same shape, and is made of material of the same density, it must be eight times as heavy. It is probably not obvious why I have chosen 1.4 for times, but I will explain after a few more paragraphs.

Different systems can move in dynamically similar fashion, only if certain conditions are satisfied. What those conditions are depends on what forces are important. In running and jumping, gravity is important, so we will discover the conditions for dynamic similarity with gravity important. It can be obtained in various ways: here is one of them.

Movements under gravity generally involve potential energy being converted to kinetic energy, and vice versa. For example, energy is swapped back and forth between the kinetic and potential forms in a swinging pendulum. A ball thrown upwards loses kinetic energy as it gains potential energy. Therefore, dynamically similar movement is possible only if the systems in question have the same ratio of kinetic to potential energy. A body of mass m travelling with velocity v has kinetic energy $\frac{1}{2}mv^2$. If it is at a height h above the ground it has potential energy mgh. The ratio of these is $v^2/2gh$. Therefore, dynamically similar motion is possible only if the systems have equal values of v^2/gh. This number, which is dimensionless, is called the Froude number.

Remember that dynamic similarity requires all lengths to be in the same proportion. Therefore, the quantity h in the Froude number need not be a height but may be any linear dimension, provided the same dimension is used for all the motions being compared. Also, times and therefore velocities are required to be in the same proportion. Therefore, v may be any characteristic velocity.

Now I will explain the factor 1.4, which I introduced mysteriously a few paragraphs back. I suggested that an animal twice as long as another might take strides twice as long, and that each stride might occupy 1.4 times as much time, when their movements were dynamically similar. The factor 1.4 was an approximation for the square root of two. Velocity is stride length divided by stride time, so the larger animal would travel $2/1.4 = 1.4$ times as fast as the small one. Its Froude number v^2/gh would be $1.4^2/2 = 1$ times the Froude number for the small one, which it would have to be for dynamic similarity to be possible.

The dynamic similarity hypothesis was first introduced in a discussion of dinosaur gaits (Alexander, 1976) but has been developed most fully for quadrupedal mammals (Alexander & Jayes, 1983). It says that terrestrial animals tend to walk and run in dynamically similar fashion, when travelling with equal Froude numbers v^2/gh. The speed v conventionally used for calculating the Froude number is the mean forward speed (averaged over a complete stride). The height h is the height of the hip joint from the ground when standing. If animals of different sizes were geometrically similar in all respects, any linear dimension could be used for h (for example, the width of the third premolar tooth). In practice it seems sensible to use a length that is functionally related to gait.

The dynamic similarity hypothesis predicts that different mammals will use the same gait, at any particular Froude number. In confirmation of this, most quadrupedal mammals change from walking to trotting at a Froude number of about 0.5 and from trotting to galloping at a Froude number of about 2.5 (Alexander & Jayes, 1983). Bipeds also change from walking to running or hopping at a Froude number of about 0.5 (Hayes & Alexander, 1983). The hypothesis also predicts that different animals travelling at the same Froude number will have their feet on the ground for the same fraction of each stride, and exert forces that are equal multiples of body weight. Further, their stride lengths will be in proportion to their leg lengths. All these predictions are quite well confirmed, but there is a complication in the case of stride length. It seems necessary to distinguish three groups of quadrupedal mammals: cursorial mammals such as dogs and horses which move on fairly straight legs, non-cursorial mammals such as rats and ferrets which move on bent legs, and primates. Fig. 9.3 shows relative stride length (stride length divided by hip height) plotted against Froude number. It shows that at any given Froude number, all cursorial mammals, tend to use approximately equal relative stride lengths, and all non-cursorial mammals (with the possible exception of coypu) tend to use approximately equal relative stride lengths. (The coypu, *Myocastor*, were far from tame and may not have been behaving normally.) However, at any given Froude number, non-cursorial mammals use longer relative stride lengths than cursorial ones, and primates (Alexander & Maloiy, 1984) use still longer ones.

There is a good theoretical reason for expecting different animals to move in dynamically similar fashion. Work is force multiplied by distance. In dynamically similar motions of different animals, all

Fig. 9.3. Graph on logarithmic coordinates showing relative stride length plotted against Froude number. Filled symbols are used for cursorial mammals and hollow ones for non-cursorial ones, as follows: ●, dog; ■, sheep; ◆, camel; ▲, ▼, rhinoceros species; ★, cat; small dots, horse; ○, ferret; □, rat; △, jird; ▽, coypu. From R. McN. Alexander & A. S. Jayes (1983). *J. Zool.*, **201**, 135–52.

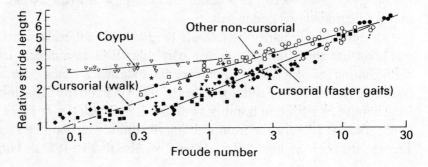

forces are proportional to body weight and all distances to stride length. Therefore, the work done by the muscles, during a stride, is proportional to (body weight) × (stride length). Consequently, animals moving in dynamically similar fashion have equal costs of transport (work performed in moving unit weight of animal a unit distance). Suppose, as seems likely, that an animal adjusts its gait to minimize cost of transport. Another animal minimizing its cost of transport at the same Froude number must move in dynamically similar fashion.

In running, elastic forces are important as well as gravity. True dynamic similarity requires a second condition, as well as equal Froude numbers. At corresponding stages of the stride, corresponding parts of the body must show similar elastic deformations.

In dynamically similar motions, all forces are proportional to body weight mg and all linear dimensions to hip height h. Therefore, stresses (force/area) are proportional to mg/h^2 and strains to mg/Eh^2 where E is the appropriate Young's modulus. The mass m of the body is proportional to ϱh^3, where ϱ is the density. Hence, strain is proportional to $\varrho h^3 g/Eh^2 = \varrho gh/E$. Dynamic similarity is possible only if the animals have equal values of the number ϱghE. This is another dimensionless number, like the Froude number.

Animals made of the same materials have equal values of ϱ and E. If they live on earth, they experience the same value of g. Therefore, to have equal values of $\varrho gh/E$, they must have equal values of h. Strict dynamic similarity is not possible for animals of different size made of the same materials, if gravity and elastic forces are both important.

9.3 Compromises

The dynamic similarity hypothesis may still be useful, even for discussions of running. I have shown that it is rather successful, in predicting details of gait. How can this be?

The paths of points on the body depend far more on the lengths of body segments, than on their diameters. The fluctuations of kinetic and potential energy that occur in locomotion would be changed very little if every segment were shrunk down to an infinitesimally thin line of the same length and mass (like a stick figure). Therefore, optimum gaits are altered very little by differences in diameter. Animals with similar proportions in respect of the lengths of their body segments can be expected to move in dynamically similar fashion, even if their diameters are out of proportion. A slender gazelle and a stout buffalo can be expected to use the same gait, at any given Froude number, with equal relative stride lengths. Fig. 9.3 shows that even animals as different in proportions as camels and rhinoceroses use about the same relative stride length, at any Froude number.

Elastically similar animals bend in similar fashion under the weights of their bodies: displacements such as s (Fig. 9.1(a)) are proportional to segment lengths. Animals moving in dynamically similar fashion exert forces proportional to body weight. Therefore, elastically similar animals suffer similar elastic deformations if they move in ways that are dynamically similar (apart from the discrepancy of diameters). Strain energy stored in their tendons will account for the same proportion of work done in the stride, in each animal.

The combination of elastic similarity with (imperfect) dynamic similarity is a compromise that could overcome the incompatibility of elastic mechanisms with perfect dynamic similarity. However, there are other compromises to be made. When geometrically similar animals move in dynamically similar fashion, forces are proportional to body mass. Cross-sectional areas are proportional to (length)^2 or (body mass)$^{0.67}$. Therefore, stresses are proportional to (body mass)$^{0.33}$. When elastically similar animals move in dynamically similar fashion, muscle and tendon forces are proportional to (weight $\times$ length)/(diameter)^3 (equation 9.7) and therefore to (mass) $\times$ (mass)$^{0.25}$/(mass)$^{1.125}$ (proportionalities 9.14), or to (mass)$^{0.125}$. Both for geometrically similar animals and for elastically similar ones, dynamically similar motion means larger stresses in larger animals. The effect is more severe in the case of geometric similarity but is present in both cases.

It is presumably desirable to minimize energy consumption in running, at any particular speed. However, stresses must be kept below the limits set by the strengths of muscle, bone and tendon. It may be necessary to compromise between these requirements. An animal running near its maximum speed may have to use a gait that would not be the most economical, if it could stand larger stresses. Consequently, the optimum gaits for large and small animals may not be precisely dynamically similar, even when both run at the same Froude number. For example, buffaloes (*Syncerus caffer*) and gazelles (*Gazella thomsonii*) reached maximum Froude numbers of about 4 and 40, respectively, when chased by a vehicle over level grassland (data of Alexander, Langman & Jayes, 1977). At a Froude number of 4, a buffalo was near maximum speed and might have to use stress-reducing strategies, but a gazelle would be far below maximum speed and could concentrate on minimizing energy costs.

The most obvious difference between the gaits of large mammals and small ones is that large (cursorial) mammals run with their legs straighter than small (non-cursorial) ones, and have shorter relative stride lengths at any particular Froude number (Fig. 9.3). Almost all mammals heavier than 5 kg use the straight-legged (cursorial) style of running and almost all lighter than 1 kg use the bent-legged (non-cursorial) style (Jenkins, 1971). Cursorial and non-cursorial mammals have their feet on the ground for approximately equal fractions of the stride, at any particular Froude number, so the forces

Fig. 9.4. Diagrams of the skeleton of the fore legs of a 0.15 kg ground squirrel (*Spermophilus*) and a 270 kg horse, at corresponding stages of their strides. The large arrows represent the ground force F_g and the force exerted by the triceps muscle F_m. From A. A. Biewener (1983). *J. exp. Biol.*, **105**, 147–71.

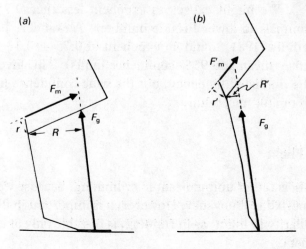

on them are approximately proportional to body mass (Alexander & Jayes, 1983). However, the straighter legs of cursorial mammals reduce the moment arms of the ground force about the joints, reducing the forces required in the muscles. Fig. 9.4 represents the forelimbs of a ground squirrel and a horse, at corresponding stages of their strides. F_g is the ground force and F_m is the force exerted by the triceps, the extensor muscle of the elbow. By taking moments about the elbow

$$F_m \approx (R/r)F_g \qquad (9.15)$$

(This equation is only approximate because the weight and inertia of the foot have been ignored). Plainly, F_m is a much smaller multiple of F_g for the horse than for the squirrel. The shorter relative stride lengths of cursorial mammals are a consequence of their straighter legs: their legs are shorter, for a given hip height, than the bent legs of non-cursorial mammals, and cannot reach as far forward and back at the beginning and end of a step.

In addition to such departures from dynamic similarity, some departure from elastic similarity may be desirable, because different tendon stiffnesses are optimal for different Froude numbers (section 3.4). Large animals might be expected to evolve tendons adapted to lower Froude numbers than smaller animals: these would be less stiff than required for elastic similarity.

It seems simplest to look for evidence of this in the Bovidae, because their skeletons are close to elastic similarity. Elastically similar tendons would have cross-sectional areas proportional to (body mass)$^{0.75}$. We might expect an exponent less than 0.75, to adapt larger animals to lower Froude numbers. Alexander, Jayes, Maloiy & Wathuta (1981) found an exponent of 0.75 ± 0.13 for a range of Bovidae (mean and 95 % confidence limits). This gives no evidence of the predicted tendency, but the wide confidence limits leave a lot of room for uncertainty.

9.4 Flight

Gravity is rather unimportant in swimming, because weight is largely supported by buoyancy. However, it is important in flight. Dynamic similarity in flight, as in running, is possible only at equal Froude numbers.

For example, consider level flight, when the lift on the wings must equal the weight mg of the body. The lift is also given by an expression involving air density ϱ, wing area A, speed v and a quantity C_L called the lift coefficient

$$\text{lift} = mg = \tfrac{1}{2}\varrho A v^2 C_L$$
$$C_L = 2mg/\varrho A v^2 \tag{9.16}$$

For geometrically similar animals, m is proportional to (length)^3 and A to (length)^2. Air density ϱ is almost constant. For equal lift coefficients, animals must fly with equal values of lg/v^2 and so of the Froude number v^2/gl, where l represents length.

The viscosity of the air is important in flight, so true dynamic similarity (of the movement of the air as well as of the animals) requires a second condition: the Reynolds numbers must be equal, as well as the Froude numbers. The two conditions could not be met simultaneously, for animals of different sizes in the same fluid, but this has little practical importance because lift and drag coefficients change little with changing Reynolds numbers, in the range of Reynolds numbers used by flying animals.

It was shown in Chapter 4 that elastic mechanisms are important in the flight of some animals, introducing a third condition for dynamic similarity (section 9.2). It might seem that this condition could be satisfied by a compromise between dynamic and elastic similarity, as in running. This might involve making wing lengths proportional to (body mass)$^{0.25}$ and wing chords (widths) proportional to (body mass)$^{0.375}$. However, this would not work well because aspect ratio (the ratio of wing length to chord) has a great effect on flight performance. The stick figure analogy that was used to justify the compromise between dynamic and elastic similarity for running cannot be applied to flight. Birds and insects actually tend to have wing lengths about proportional to (body mass)$^{0.4}$ and chords about proportional to (body mass)$^{0.3}$ (Alexander, 1982). I know no group of flying animals whose proportions tend to elastic similarity.

References

Alexander, R. McN. (1962). Visco-elastic properties of the body wall of sea anemones. *Journal of experimental Biology*, **39**, 373–86.

Alexander, R. McN. (1966). Rubber-like properties of the inner hinge-ligament of Pectinidae. *Journal of experimental Biology*, **44**, 119–30.

Alexander, R. McN. (1976). Estimates of speeds of dinosaurs. *Nature*, **261**, 129–30.

Alexander, R. McN. (1979). *The Invertebrates*. Cambridge: Cambridge University Press.

Alexander, R. McN. (1980). Optimum walking techniques for quadrupeds and bipeds. *Journal of Zoology*, **192**, 97–117.

Alexander, R. McN. (1981). Factors of safety in the structure of animals. *Science Progress*, **67**, 109–30.

Alexander, R. McN. (1982). Size, shape and structure for running and flight. In *A Companion to Animal Physiology*, ed. C. R. Taylor, K. Johansen & L. Bolis, pp. 309–24. Cambridge: Cambridge University Press.

Alexander, R. McN. (1983). *Animal Mechanics* (2nd edn). Oxford: Blackwell Scientific.

Alexander, R. McN. (1984). Elastic energy stores in running vertebrates. *American Zoologist*, **24**, 85–94.

Alexander, R. McN. (1985). The maximum forces exerted by animals. *Journal of experimental Biology*, **115**, 231–8.

Alexander, R. McN. (1986). Making headway in Africa. *Nature*, **319**, 623–4.

Alexander, R. McN. (1987). Bending of cylindrical animals with helical fibres in their skin or cuticle. *Journal of theoretical Biology*, **124**, 97–110.

Alexander, R. McN. & Bennet-Clark, H. C. (1977). Storage of elastic strain energy in muscle and other tissues. *Nature*, **265**, 114–17.

Alexander, R. McN., Bennett, M. B. & Ker, R. F. (1986). Mechanical properties and functions of the paw pads of some mammals. *Journal of Zoology*, A, **209**, 405–19.

Alexander, R. McN. & Dimery, N. J. (1985*a*). The significance of sesamoids and retro-articular processes for the mechanics of joints. *Journal of Zoology*, A, **205**, 357–71.

Alexander, R. McN. & Dimery, N. J. (1985*b*). Elastic properties of the fore foot of the donkey (*Equus asinus*). *Journal of Zoology*, A, **205**, 511–24.

Alexander, R. McN., Dimery, N. J. & Ker, R. F. (1985). Elastic structures in the back and their role in galloping in some mammals. *Journal of Zoology*, A, **207**, 467–82.

Alexander, R. McN. & Jayes, A. S. (1983). A dynamic similarity hypothesis for the gaits of quadrupedal mammals. *Journal of Zoology*, **201**, 135–52.

Alexander, R. McN., Jayes, A. S., Maloiy, G. M. O. & Wathuta, E. M. (1979). Allometry of the limb bones of mammals from shrews (*Sorex*) to elephant (*Loxodonta*). *Journal of Zoology*, **190**, 155–92.

Alexander, R. McN., Jayes, A. S., Maloiy, G. M. O. & Wathuta, E. M. (1981). Allometry of the leg muscles of mammals. *Journal of Zoology*, **194**, 539–52.

Alexander, R. McN., Langman, V. A. & Jayes, A. S. (1977). Fast locomotion of some African ungulates. *Journal of Zoology*, **183**, 291–300.

Alexander, R. McN. & Maloiy, G. M. O. (1984). Stride lengths and stride frequencies of primates. *Journal of Zoology*, **202**, 577–82.

Alexander, R. McN., Maloiy, G. M. O., Ker, R. F., Jayes, A. S. & Warui, C. N. (1982). The role of tendon elasticity in the locomotion of the camel (*Camelus dromedarius*). *Journal of Zoology*, **198**, 293–313.

Alexander, R. McN., Maloiy, G. M. O., Njau, R. & Jayes, A. S. (1979). Mechanics of running of the ostrich (*Struthio camelus*). *Journal of Zoology*, **187**, 169–78.

Alexander, R. McN. & Vernon, A. (1975). Mechanics of hopping by kangaroos (Macropodidae). *Journal of Zoology*, **177**, 265–303.

Batham, E. J. & Pantin, C. F. A. (1950). Muscular and hydrostatic action in the sea anemone, *Metridium senile* (L.). *Journal of experimental Biology*, **27**, 264–89.

Bennet-Clark, H. C. (1975). The energetics of the jump of the locust, *Schistocerca gregaria*. *Journal of experimental Biology*, **63**, 53–83.

Bennet-Clark, H. C. & Alder, G. M. (1979). The effect of air resistance on the jumping performance of insects. *Journal of experimental Biology*, **82**, 105–21.

Bennet-Clark, H. C. & Lucey, E. C. A. (1967). The jump of the flea: a study of the energetics and a model of the mechanism. *Journal of experimental Biology*, **47**, 59–76.

Bennet, M. B. & Alexander, R. McN. (1987). Properties and functions of extensible ligaments in the necks of turkeys (*Meleagris gallopavo*) and other birds. *Journal of Zoology*, **212**, 275–81.

Bennett, M. B., Ker, R. F. & Alexander, R. McN. (1987). Elastic properties of structures in the tails of cetaceans (*Phocaena* and *Lagenorhynchus*) and their effect on the energy cost of swimming. *Journal of Zoology*, **211**, 177–92.

Bennett, M. B., Ker, R. F., Dimery, N. J. & Alexander, R. McN. (1986). Mechanical properties of various mammalian tendons. *Journal of Zoology*, *A*, **209**, 537–48.

Biewener, A. A. (1983). Locomotory stresses in the limb bones of two small mammals: the ground squirrel and the chipmunk. *Journal of experimental Biology*, **105**, 147–71.

Biewener, A., Alexander, R. McN. & Heglund, N. C. (1981). Elastic energy storage in the hopping of kangaroo rats (*Dipodomys spectabilis*). *Journal of Zoology*, **195**, 369–83.

Boettiger, E. G. & Furshpan, E. (1952). The mechanics of flight movements in Diptera. *Biological Bulletin*, **102**, 200–11.

Bramble, D. M. & Carrier, D. R. (1983). Running and breathing in mammals. *Science*, **219**, 251–6.

Brown, R. H. J. (1963). The flight of birds. *Biological Reviews*, **38**, 460–89.

Brown, T. I. H., Rack, P. M. H. & Ross, H. F. (1982). Forces generated at the thumb interphalangeal joint during imposed sinusoidal movements. *Journal of Physiology*, **332**, 69–85.

Casey, T. M. (1981). A comparison of mechanical and energetic estimates of flight costs for hovering sphinx moths. *Journal of experimental Biology*, **91**, 117–29.

Cavagna, G. A. (1975). Force platforms as ergometers. *Journal of applied Physiology*, **39**, 174–9.

Cavagna, G. A., Heglund, N. C. & Taylor, C. R. (1977). Mechanical work in terrestrial locomotion: two basic mechanisms for minimizing energy expenditure. *American Journal of Physiology*, **233**, R243–R261.

Cavagna, G. A., Saibene, F. B. & Margaria, R. (1964). Mechanical work in running. *Journal of applied Physiology*, **19**, 249–56.

Cavanagh, P. R. & Lafortune, M. A. (1980). Ground reaction forces in distance running. *Journal of Biomechanics*, **13**, 397–406.

Clark, J. & Alexander, R. McN. (1975). Mechanics of running by quail (*Coturnix*). *Journal of Zoology*, **176**, 87–113.

Clark, R. B. (1964). *Dynamics in Metazoan Evolution*. Oxford: Oxford University Press.

Clark, R. B. & Cowey, J. B. (1958). Factors controlling the change of shape of certain nemertean and turbellarian worms. *Journal of experimental Biology*, **35**, 731–48.

Close, R. I. (1972). Dynamic properties of mammalian skeletal muscles. *Physiological Reviews*, **52**, 129–97.

Crawford, E. C. (1962). Mechanical aspects of panting in dogs. *Journal of applied Physiology*, **17**, 249–51.

Currey, J. D. (1984). *The Mechanical Adaptations of Bones*. Princeton: Princeton University Press.

Cutts, A. (1986). Sarcomere length changes in the wing muscles during the wing beat cycle of two bird species. *Journal of Zoology*, *A*, **209**, 183–5.

Diamant, J., Keller, A., Baer, E., Litt, M. & Arridge, R. G. C. (1972). Collagen: ultrastructure and its relation to mechanical properties as a function of ageing. *Proceedings of the Royal Society*, *B*, **180**, 293–315.

Dickinson, J. A., Cook, S. D. & Leinhardt, T. M. (1985). The measurement of shock waves following heel strike while running. *Journal of Biomechanics*, **18**, 415–22.

Dimery, N. J. & Alexander, R. McN. (1985). Elastic properties of the hind foot of the donkey, *Equus asinus*. *Journal of Zoology*, *A*, **207**, 9–20.

Dimery, N. J., Alexander, R. McN. & Deyst, K. A. (1985). Mechanics of the ligamentum nuchae of some artiodactyls. *Journal of Zoology*, *A*, **206**, 341–51.

Dimery, N. J., Alexander, R. McN. & Ker, R. F. (1986). Elastic extensions of leg tendons in the locomotion of horses (*Equus caballus*). *Journal of Zoology*, *A*, **210**, 415–25.

Dimery, N. J., Ker, R. F. & Alexander, R. McN. (1986). Elastic properties of the feet of deer (Cervidae). *Journal of Zoology*, *A*, **208**, 161–9.

Economos, A. C. (1982). On the origin of biological similarity. *Journal of theoretical Biology*, **94**, 25–60.

Ellington, C. P. (1984*a*). The aerodynamics of hovering insect flight. III. Kinematics. *Philosophical Transactions of the Royal Society*, *B*, **305**, 41–78.

Ellington, C. P. (1984*b*). The aerodynamics of hovering insect flight. V. A vortex theory. *Philosophical Transactions of the Royal Society*, *B*, **305**, 115–44.

Ellington, C. P. (1984*c*). The aerodynamics of hovering insect flight. VI. Lift and power requirements. *Philosophical Transactions of the Royal Society*, *B*, **305**, 145–81.

Ennos, A. R. (1987). A comparative study of the flight mechanism of Diptera. *Journal of experimental Biology*, **127**, 355–72.

Evans, M. E. G. (1972). The jump of the click beetle (Coleoptera, Elateridae) – a preliminary study. *Journal of Zoology*, **167**, 319–36.

Evans, M. E. G. (1973). The jump of the click beetle (Coleoptera: Elateridae) – energetics and mechanics. *Journal of Zoology*, **169**, 181–94.

Flitney, F. W. & Hirst, D. G. (1978). Cross-bridge detachment and sarcomere 'give' during stretch of active frog's muscle. *Journal of Physiology*, **276**, 449–65.

Gambaryan, P. P. (1974). *How Mammals Run*. New York: Wiley.

Gordon, A. M., Huxley, A. F. & Julian, F. J. (1966). The variation in isometric tension with sarcomere length in vetebrate muscle fibres. *Journal of Physiology*, **184**, 170–92.

Gosline, J. M. (1971). Connective tissue mechanics of *Metridium senile*. II. Viscoelastic properties and a macromolecular model. *Journal of experimental Biology*, **55**, 775–95.

Gosline, J. M. (1980). The elastic properties of rubber-like proteins and highly extensible tissues. In *The Mechanical Properties of Biological Materials*, Symposia of the Society for experimental Biology 34, ed. J. F. V. Vincent & J. D. Currey, pp. 331–57. Cambridge: Cambridge University Press.

Hall-Craggs, E. C. B. (1965). An analysis of the jump of the Lesser galago (*Galago senegalensis*). *Journal of Zoology*, **147**, 20–9.

Harris, J. E. & Crofton, H. D. (1957). Structure and function in nematodes: internal pressure and cuticular structure in *Ascaris*. *Journal of experimental Biology*, **34**, 116–30.

Hayes, G. & Alexander, R. McN. (1983). The hopping gaits of crows (Corvidae) and other bipeds. *Journal of Zoology*, **200**, 205–13.

Hebrank, M. R. (1980). Mechanical properties and locomotor function of eel skin. *Biological Bulletin*, **158**, 58–68.

Heglund, N. C., Fedak, M. A., Taylor, C. R. & Cavagna, G. A. (1982). Energetics and mechanics of terrestrial locomotion. IV. Total mechanical energy changes as a function of speed and body size in birds and mammals. *Journal of experimental Biology*, **97**, 57–66.

Heglund, N. C., Taylor, C. R. & McMahon, T. A. (1974). Scaling stride frequency and gait to animal size: mice to horses. *Science*, **186**, 1112–13.

Heitler, W. J. (1974). The locust jump. Specialisations of the metathoracic femoral–tibial joint. *Journal of comparative Physiology*, **89**, 93–104.

Hokkanen, J. E. I. (1986). Notes concerning elastic similarity. *Journal of theoretical Biology*, **120**, 499–501.

Hoyt, D. F. & Taylor, C. R. (1981). Gait and the energetics of locomotion in horses. *Nature*, **292**, 239–40.

Huxley, A. F. & Simmons, R. M. (1971). Mechanical properties of the cross bridges of frog striated muscle. *Journal of Physiology*, **218**, 59P–60P.

Jayes, A. S. & Alexander, R. McN. (1978). Mechanics of locomotion of dogs (*Canis familiaris*) and sheep (*Ovis aries*). *Journal of Zoology*, **185**, 289–308.

Jenkins, F. A. (1971). Limb posture and locomotion in the Virginia opossum (*Didelphis marsupialis*) and in other non-cursorial mammals. *Journal of Zoology*, **165**, 303–15.

Jensen, M. & Weis-Fogh, T. (1962). Biology and physics of locust flight. V. Strength and elasticity of locust cuticle. *Philosophical Transactions of the Royal Society*, B, **245**, 137–69.

Kahler, G. A., Fisher, F. M. & Sass, R. L. (1976). The chemical composition and mechanical properties of the hinge ligament in bivalve molluscs. *Biological Bulletin*, **151**, 161–81.

Ker, R. F. (1980). Small-scale tensile tests. In *The Mechanical Properties of Biological Materials*, Symposia of the Society for experimental Biology 34, ed. J. F. V. Vincent & J. D. Currey, pp. 487–9. Cambridge: Cambridge University Press.

Ker, R. F. (1981). Dynamic tensile properties of the plantaris tendon of sheep. (*Ovis aries*). *Journal of experimental Biology*, **93**, 283–302.

Ker, R. F., Bennett, M. B., Bibby, S. R., Kester, R. C. & Alexander, R. McN. (1987). The spring in the arch of the human foot. *Nature*, **325**, 147–9.

Ker, R. F., Dimery, N. J. & Alexander, R. McN. (1986). The role of tendon elasticity in hopping in a wallaby (*Macropus rufogriseus*). *Journal of Zoology, A*, **208**, 417–28.

Koehl, M. A. R. (1977). Mechanical diversity of connective tissue of the body wall of sea anemones. *Journal of experimental Biology*, **69**, 107–25.

La Barbera, M. (1983). Why the wheels won't go. *American Naturalist*, **121**, 395–408.

Lang, T. G. & Daybell, D. A. (1963). Porpoise performance tests in a sea-water tank. *United States Naval Ordnance Test Station Technical Publication* no. 3063. NAVWEPS Report 8060, 1–50.

Light, L. H., McLellan, G. E. & Klenerman, L. (1980). Skeletal transients on heel strike in normal walking with different footwear. *Journal of Biomechanics*, **13**, 477–80.

Lochner, F. K., Milne, D. W., Mills, E. J. & Groom, J. J. (1980). In vivo and in vitro measurement of tendon strain in the horse. *American Journal of veterinary Research*, **41**, 1929–37.

Machin, K. E. & Pringle, J. W. S. (1959). The physiology of insect fibrillar muscle. II. Mechanical properties of a beetle flight muscle. *Proceedings of the Royal Society, B*, **151**, 204–25.

McMahon, T. A. (1973). Size and shape in biology. *Science*, **179**, 1201–4.

McMahon, T. A. (1975). Using body size to understand the structural design of animals: quadrupedal locomotion. *Journal of applied Physiology*, **39**, 619–27.

Maloiy, G. M. O., Heglund, N. C., Prager, L. M., Cavagna, G. A. & Taylor, C. R. (1986). Energetic cost of carrying loads: have African women discovered an economic way? *Nature*, **319**, 668–9.

Miyan, J. A. & Ewing, A. W. (1985a). How Diptera move their wings: a re-examination of the wing base articulation and muscle systems concerned with flight. *Philosophical Transactions of the Royal Society, B*, **311**, 271–302.

Miyan, J. A. & Ewing, A. W. (1985b). Is the 'click' mechanism of dipteran flight an artefact of CCl$_4$ anaesthesia? *Journal of experimental Biology*, **116**, 313–22.

Moore, J. D. & Trueman, E. R. (1971). Swimming of the scallop, *Chlamys opercularis* (L). *Journal of experimental marine Biology and Ecology*, **6**, 179–85.

Morgan, D. L., Proske, U. & Warren, D. (1978). Measurements of muscle stiffness and the mechanism of elastic storage in hopping kangaroos. *Journal of Physiology*, **282**, 253–61.

Mullins, L. (1980). Theories of rubber-like elasticity and the behaviour of filled rubber. In *The Mechanical Properties of Biological Materials*, Symposia of the Society for experimental Biology, 34, ed. J. F. V. Vincent & J. D. Currey, pp. 273–88. Cambridge: Cambridge University Press.

Pennycuick, C. J. & Lock, A. (1976). Elastic energy storage in primary feather shafts. *Journal of experimental Biology*, **64**, 677–89.

Pugh, L. G. C. E. (1974). The relation of oxygen intake and speed in competition cycling and comparative observations on the bicycle ergometer. *Journal of Physiology*, **241**, 795–808.

Rack, P. M. H. & Ross, H. F. (1984). The tendon of flexor pollicis longus: its effects on the muscular control of force and position at the human thumb. *Journal of Physiology*, **351**, 99–110.

Rack, P. M. H., Ross, H. F., Thilmann, A. F. & Walters, D. K. W. (1983). Reflex responses at the human ankle: the importance of tendon compliance. *Journal of Physiology*, **344**, 503–24.

Rack, P. M. H. & Westbury, D. R. (1974). The short range stiffness of active mammalian muscle and its effect on mechanical properties. *Journal of Physiology*, **240**, 331–50.

Rashevsky, N. (1960). *Mathematical Biophysics*. New York: Dover.

Rasmussen, S., Chan, A. K. & Goslow, G. E. (1978). The cat step cycle: electromyographic patterns for hind limb muscles during posture and unrestrained locomotion. *Journal of Morphology*, **155**, 253–70.

Schmidt-Nielsen, K. (1984). *Scaling. Why is animal size so important?* Cambridge: Cambridge University Press.

Todd, D. J. (1985). *Walking Machines*. London: Kogan Page.

Trueman, E. R. (1953). Observations on certain mechanical properties of the ligament of *Pecten*. *Journal of experimental Biology*, **30**, 453–67.

Usherwood, P. N. R. (ed.) (1975). *Insect Muscle*. New York: Academic Press.

Videler, J. J. (1975). On the interrelationships between morphology and movement in the tail of the cichlid fish *Tilapia nilotica* (L.). *Netherlands Journal of Zoology*, **25**, 143–94.

Vincent, J. F. V. (1980). Insect cuticle: a paradigm for natural composites. In *The Mechanical Properties of Biological Materials*, Symposia of the Society for experimental Biology 34, ed. J. F. V. Vincent & J. D. Currey, pp. 183–210. Cambridge: Cambridge University Press.

Vincent, J. F. V. & Currey, J. D. (ed.) (1980). *The Mechanical Properties of Biological Materials*. Symposia of the Society for experimental Biology 34, 513 pp. Cambridge: Cambridge University Press.

Wainwright, S. A., Biggs, W. D., Currey, J. D. & Gosline, J. M. (1976). *Mechanical Design in Organisms*. London: Edward Arnold.

Wainwright, S. A., Pabst, D. A. & Brodie, P. F. (1985). Form and possible function of the collagen layer underlying cetacean blubber. *American Zoologist*, 25, 146A.

Wainwright, S. A., Vosburgh, F. & Hebrank, J. H. (1978). Shark skin: function in locomotion. *Science*, 202, 747–9.

Weis-Fogh, T. (1960). A rubber-like protein in insect cuticle. *Journal of experimental Biology*, 37, 889–906.

Weis-Fogh, T. (1961a). Power in flapping flight. In *The Cell and the Organism*, ed. J. A. Ramsay & V. B. Wigglesworth, pp. 283–300. Cambridge: Cambridge University Press.

Weis-Fogh, T. (1961b). Thermodynamic properties of resilin, a rubber-like protein. *Journal of molecular Biology*, 3, 520–31.

Weis-Fogh, T. (1972). Energetics of hovering flight in hummingbirds and *Drosophila*. *Journal of experimental Biology*, 56, 79–104.

Weis-Fogh, T. (1973). Quick estimates of flight fitness in hovering animals, including novel mechanisms for lift production. *Journal of experimental Biology*, 59, 169–230.

Wong, J. Y. (1978). *Theory of Ground Vehicles*. New York: Wiley.

Index